MODERN ALGEBRA

DPH MATHEMATICS SERIES

MODERN ALGEBRA

(For M.A., M.Sc., I.A.S., P.C.S.)

By

A.K. Sharma

DISCOVERY PUBLISHING HOUSE
NEW DELHI-110002

Reprinted - 2019

First Published - 2004

ISBN: 978-81-7141-888-6

Modern Algebra

Published by:

DISCOVERY PUBLISHING HOUSE PVT. LTD.
4383/4B, Ansari Road, Darya Ganj
New Delhi-110 002 (India)
Phone: +91-11-23279245, 23253475; 43596065
E-mail: discoverybooksindia@gmail.com
discoverypublishinghouse@gmail.com
web: www.discoverypublishinggroup.com

Printed at:
Infinity Imaging Systems
Delhi

Preface

This book "Modern Algebra" has been specially written to meet the requirements M.A./M.Sc., students of all Indian Universities.

The subject matter has been discussed in such a simple way that the students will find no difficulty to understand it. The proof of various theorems and examples have been given with minute details. Each chapter of this book contains complete theory and large number of solved examples. Sufficient problems have also been selected from various Indian Universities and competitive examination.

I hope that this book warmly received by the students and teachers.

The author wish to express their thanks to the publisher M/s Discovery Publishing House, New Delhi for bringing out this book in the present nice form.

A.K. Sharma

CONTENTS

Rings of Residue Classes, Homomorphism of Rings, Kernel of a Ring Homomorphism, Maximal Ideal, Some More Results on Ideals, Prime Ideals, Euclidean Rings or Euclidean Domains, Properties of Euclidean Rings, Ideal Generated by a Given Subset of a Ring, Principal Ideal Ring, Divisibility in an Integral Domain, Proper and Improper Divisors, Relatively Prime Elements, Degree of Polynomial, Degree of the Sum and the Product of Two Polynomials, Ring of Polynomials, Polynomials Over a Integral Domain.

4. **Logics**

Introduction, Statement, Truth Value of a Statement, Equivalent Statements, Simple Statement, Compound Statement, Logic, Logical Connectives, Argument, Laws of Algebra of Proposition, Tautology and Fallacy, Mathematical System, Proofs, Methods of Proof, Normal Forms, Propositions Over a Universe, Quantifiers,

1

RINGS

1.1 INTRODUCTION

In this chapter we studied an algebraic system known as a group in our study of groups. We considered sets which were closed w.r.t. a single operation. Here we shall study an important type of algebraic structure with two binary operation known as a Rings. A Rings has its origin in the set of integrals which is closed w.r.t. both the operation of ordinary addition and ordinary multiplication.

Definition: *Let R is a non-empty set equipped with two binary operations called addition and denoted by '+' and '.' respectively i.e., for all a, b, ∈ R we have $a + b \in R$ and $a.b \in R$. Then this algebraic structure (R, +, .) is called a right if the following postulates are satisfied:*

1. *Addition of associative, i.e.,*

   $(a + b) + c = a + (b + c) \; \forall \; a, b, c \in R.$

2. *Addition is commutative, i.e.,* $a + b = b + a \; \forall \; a, b, \in R.$

3. *There exists an element denoted by 0 in R such that*

   $0 + a = a \; \forall \; a \in R.$

4. *To each element a in R there exists an element –a in R such that*

   $(-a) + a = 0$

5. *Multiplication is associative, i.e.,*

   $a. (b, c) = (a.b) . c \; \forall \; a, b, c \in R.$

6. Multiplication is distributive with respect to addition, *i.e.*, for all a, b, c in R,

$$\left.\begin{aligned} a.(b+c) &= a.b + a.c. \\ \text{and} \quad (b+c)a &= b.a + c.a \end{aligned}\right\}$$

Sine addition is commutative in R, therefore we shall have $0 \in R$ such that $0 + a = a + 0 \; \forall \; a \in R$.

Also if $a \in R$. Then we shall have $(-a) + a = 0 = a + (-a)$. Thus, R will be an abelian group w.r.t. addition. The element $0 \in R$ will be the additive identity it's called the zero element of the *Rings*. Since in a group the identity element is unique. Therefore every ring will passes unique zero element and it will be the additive composition. We shall always denote this element by the symbol 0.

Notes:

1. When there is no confusion about the operations on R we simply say that R is a ring
2. In a ring R, the additive identity is called the zero of the ring and the additive inverse of $a \in R$ is called the negative of a to be denoted by $-a$.

Also we define $a - b = a + (-b) \; \forall \; a, b, \in R$

The equation $a + x = b$ will have a unique solution in R and it will be $x = b - a$. Obviously $a + (b - a) = a + [b + (-a)]$

$= a + [(-a) + b] = [a + (-a)] + b = [(-a) + a] + b = 0 + b = b.$

Similarly, the equation $y + a = b$ will have a unique solution in R and it will be $y = b - a$.

Both the cancellation laws will hold good for addition in R *i.e.*, for all a, b, c in R

$$a + b = a + c \Rightarrow b = c \text{ and } b + a = c + a \Rightarrow b = c.$$

If in a ring we have $a + b = 0$, then $a = -b = i$, then $a = -b$ and $b = a$.

1.2 VARIOUS TYPE OF RING

1. **Commutative Ring:** *If in a ring R, the multiplication composition is also commutative i.e., if we have $a.\, b = b.\, a \; \forall \; a, b \in R$, then R is called a commutative ring.*
2. **A Ring without Zero Divider:** *If it is not possible to find two non-zero element of R, whose product is zero i.e., R is without zero dividers when $\{ab = 0 \Rightarrow a = 0 \; b = 0\}$*

3. **A Ring with Zero Divider:** *If it is possible to find at least two element a and b of R such that* $a \neq 0$, $b \neq 0$ *and* $ab = 0$

4. **Ring with Unity:** *If in a ring R there exists an element denoted by 1 such that* $1 . a = a = a . 1 \ \forall a \in R$, *then R is called a ring with unit element. The element* $1 \in R$ *is called the unit element of the ring. Obviously 1 is the multiplicative identity of R. Thus if a ring possesses multiplicative identity, then it is a ring with unity.*

Note: In future we shall denote the multiplication composition in a ring R not by the symbol '.' but by multiplicative notation. Thus we shall write ab in place of a. b.

1.3 ELEMENTARY PROPERTIES OF A RING

Theorem:

If R is a ring, then for all $a, b, c \in R$

1. $a0 = 0a = 0$.
2. $a(-b) = -(ab) = (-a)b$.
3. $(-a)(-b) = ab$.
4. $a(b - c) = ab - ac$
5. $(b - c)a = ba - ca$.

Proof:

1. We know that

$a0 = a(0 + 0)$ [$\because$ $0 + 0 = 0$]

$= a0 + a0$. [by left distributive law]

$\therefore 0 + a0 = a0 + a0$. [$\because$ $a0 \in R$ and $0 + a0 = a0$]

We have R is a group with respect to addition, therefore applying right cancellation law for addition in R we get $0 = a0$.

Similarly we have $0a = (0 + 0)a$

$= 0a + 0a$. [by right distributive law]

$\therefore$ $0 + 0a = 0a + 0a$ [$\because$ $0 + 0a = 0a$]

Applying right cancellation law for addition in R, we get $0 = 0a$.

2. We know that $a[(-b) + b + b] = a0$ [$\because$ $-b + b = 0$]

$\Rightarrow a(-b) + ab = 0$ [by using left distributive law and the result (1)]

$\Rightarrow$ a (–b) = – (ab), since in a ring a + b 0 $\Rightarrow$ a = – b.

Similarly we know that (–a + a) b = 0b

$\Rightarrow$ (–a) b + ab = 0 $\Rightarrow$ (–a) b = – (ab), since in a ring

a + b = 0 $\Rightarrow$ a = – b.

3. We know that (– a) (– b) = – [(– a) b], since a (– b) = – (ab) = – [–(ab)], since (– a) b = – (ab)

= ab, since R is a group with respect to addition and in a group we have

–(– a) = a.

4. We have a (b – a) = a [b + (–c)]

= ab + a (– c) [by left distributive law]

= ab + [–(ac)] [$\because$ a (–c) = – (ac)]

= ab – ac.

5. We know that

(b – c) a = [b + (–c)] a

= ba + (–c) a [by right distributive law]

= ba + [–(ca)] = ba – ca. **Hence Proved.**

1.4 INTEGRAL MULTIPLES OF THE ELEMENTS OF A RING

R is a ring and a $\in$ R. If m is a positive integer, then we define ma = a + a + a + ... upto m terms.

Also we define 0a = 0. Here 0 on the left hand side is the integer zero and 0 on the right hand side is the zero element (additive identity) of the ring.

If m is a positive integer, then –m is a negative integer. We define (–m) a = – (ma) = – (a + a + a + ... upto m terms)

= (–a) + (–a) + (–a) + ... upto m terms = m (–a).

If m and n are any integers, we can prove that

ma + na = (m + n) a, m (na)a, and (ma) (na) = (mn) a^2.

Also we can prove that for any integer m, we have

m (a + b) = ma + mb $\forall$ a, b $\in$ R.

The students should not confuse that this is the distributive law we assumed in the postulates for a ring. It is not so. Here m is an integer and a, b are elements or R.

Example 1:

The set I of all integers is a ring with respect to addition and multiplication of integers as the two ring compositions. The ring is called the ring of integers.

Solution:

As in groups, we should first prove that I is an abelian group with respect to addition of integers

1. The product of two integers is also an integer. Therefore I is closed with respect to multiplication of integers.
2. Multiplication of integers is an associative composition.
3. Multiplication of integers is distributive with respect to addition of integers *i.e.*, if a, b, c are any elements of I, then

$$a\,(b + c) = ab + ac \text{ and } (b + c)\,a = ba + ca.$$

Therefore I is a ring with respect to addition and multiplication of integers. The integer 0 is the zero element of the ring. Also the multiplicative identity exists and it is the integer 1. We have $1a = a = a1\ \forall\ a \in I$. Thus the ring of integers is a ring with unity. The integer 1 is the unit element of this ring.

Since the multiplication of integers is a commutative composition. Therefore it is also a commutative ring.

Example 2:

The set $R = \{0, 1, 2, 3, 4, 5\}$ is a commutative ring with respect to '$+_6$' and '$\times_6$' as the two ring compositions.

Solution:

We have already prove in group, we should first prove that R is an abelian group with respect to '$+_6$'.

We form a composition table for the given composition $\times_6$.

$\times_6$	0	1	2	3	4	5
0	0	0	0	0	0	0
1	0	1	2	3	4	5
2	0	2	4	0	2	4
3	0	3	0	3	0	3
4	0	4	2	0	4	2
5	0	5	4	3	2	1

From the composition table we find that R is closed with respect to the composition '$\times_6$'.

Also we know that '$\times_6$' is an associative composition in R *i.e.*, $a \times_6 (b \times_6 c) = (a \times_6 b) \times_6 c \ \forall \ a, b, c \in R$. Hence it held associate low

Further '$\times_6$' is distributive in R with respect to '$\times_6$'. If a, b, c are any elements of R, then, we have

$a \times_6 (b \times_6 c) = a \times_6 (b + c)$ $\qquad [\because \ b + c \equiv {}_6 c \pmod 6]$

= least non-negative remainder when a (b + c) is divided by 6

= least non-negative remainder when a ab + ac is divided by 6

$= (ab) +_6 (ac)$

$= (a \times_6 b) +_6 (ac)$ $\qquad [\because \ a \times {}_6 b \equiv ab \pmod 6]$

$= (a \times_6 b) +_6 (a \times_6 c)$ $\qquad [\because \ a \times {}_6 c \equiv ac \pmod 6]$

Similarly, we can prove that $(b \times_6 c) \times_6 a = (b \times_6 a) +_6 (c \times_6 a)$.

$\therefore$ R is a ring with respect to the given compositions. Since '$\times_6$' is a commutative composition in R as is clear from the composition table also, therefore R is commutative ring. Also 1 is the identity element for the composition '$\times_6$'. Therefore R is a ring with unity. The integer 0 is the zero element of the ring.

Example 3:

The set M of all $n \times n$ matrices with their elements as real numbers (rational numbers, complex numbers, integers) is a non-commutative ring with unity, with respect to addition and multiplication of matrices as the two ring compositions.

Solution:

By matrix property we know that the sum and product of two matrix is also a matrix. Therefore M is closed with respect to addition and multiplication of matrices.

Further we know that

1. $A + (B + C) = (A + B) + C \ \forall \ A, B, C \in M$, since the addition of matrices is an associative the composition
2. $A + B = B + A \ \forall \ A, \in M$ since the addition of matrices is commutative.
3. If O is the null matrix of the type $n \times n$, then $O \in M$ and we have $O + A = A \ \forall \ A \in M$. Hence O is null matrix.
4. To each matrix $A \in M$ there exists a matrix $-A \in M$ such that $(-A) + A = O$ (null matrix). Hence, $-A$ is additive inverse A.

5. (AB) C = A (BC), $\forall$ A, B, C $\in$ M, since multiplication of matrices is associative.
6. A (B + C) = AB + AC,

   and (B + C) A = BA + CA $\forall$ A, B, C $\in$ M, since matrix multiplication is distributive with respect to matrix addition.

Hence M is a ring with respect to the two given compositions. The null matrix O of the type n × n is the zero element of this ring *i.e.*, O = 0.

We know that the multiplication of matrices is not in general commutative, therefore the ring is a non-commutative ring. [n > 1]

Hence, if I be the unit of the type n × n, then I $\in$ M and we have IA = A AI " A $\in$ M. Therefore the matrix 1 is the multiplicative identity. Thus, the ring is with unity and the matrix I is the unity element of the ring *i.e.* I − 1.

Example 4:

The set R = {0, 1, 2, 3, 4, 5} is a commutative ring with respect to '$+_6$' and '$\times_6$' as the two ring compositions.

Solution:

As we have proved in groups, we should first prove that R is an abelian group with respect to '$+_6$'.

The composition table for R is as unclear.

x_6	0	1	2	3	4	5
0	0	0	0	0	0	0
1	0	1	2	3	4	5
2	0	2	4	0	2	4
3	0	3	0	3	0	3
4	0	4	2	0	4	2
5	0	5	4	3	2	1

From the composition table we have R is closed with respect to the composition '$\times_6$'.

Also we know that '$\times_6$' is an associative composition in R *i.e.*, a $\times_6$ (b $\times_6$c) (a $\times_6$b) $\times_6$c $\forall$ a, b, c $\in$ R.

Further '$\times_6$' is distributive in R with respect to '$\times_6$'. If a, b, c are any elements of R, then

$a \times_6 (b +_6 c) = a \times_6 (b + c)$ $[\because b + c \equiv b +_6 c \pmod 6]$

$=$ least non-negative remainder when $a (b + c)$ is divided by 6

$=$ least non-negative remainder when $ab + ac$ is divided by 6

$= (ab +_6 (ac)$

$= (a \times_6 b) +_6 (ac)$ $[\because a \times_6 b \equiv ab \pmod 6]$

$= (a \times_6 b) +_6 (a \times_6 c)$ $[\because a \times_6 c \equiv ac \pmod 6]$

Similarly, we can prove that $(b \times_6 c) \times_6 a = (b \times_6 a) +_6 (c \times_6 a)$.

$\therefore$ R is a ring with respect to the given compositions. Since '$\times_6$' is a commutative composition in R as is clear from the composition table also, therefore R is a commutative ring. Also 1 is a ring with unity. The integer 0 is the zero element of this ring.

Note: We see that in this ring R neither 2 nor 3 is equal to the zero element of the ring. But $2 \times_6 3 = 0$ (zero element of the ring). Thus, in a ring it is possible that the product of two non-zero elements is equal to the zero element. Also the number of elements in R is finite. Therefore this is an example of a *finite ring.*

Example 5:

The set 2I of all even integers is a commutative ring without unity, addition and multiplication of integers being the two ring compositions.

Solution:

Do yourself.

Example 6:

The set M of all $n \times n$ matrices with their elements as real numbers (rational numbers, complex numbers, integers) is a non-commutative ring with unity, with respect to addition and multiplication of matrices as the two ring compositions.

Solution:

We know that the sum and product of two $n \times n$ matrices with their elements as real numbers are again $n \times n$ matrices with their elements as real numbers. Therefore M is closed with respect to addicted and multiplication of matrices.

Here we observe that

1. $A + (B + C) = (A + B) + C \ \forall \ A, B, C \in M$, since the addition of matrices is an associate the composition

2. $A + B = B + A \; \forall \; A, \in M$ since the addition of matrices is commutative.
3. If O is the null matrix of the type $n \times n$, then $O \in M$ and we have $O + A = A \; \forall \; A \in M$.
4. To each matrix $A \in M$ there exists a matrix $-A \in M$ such that $(-A) + A = O$ (null matrix).
5. $(AB)\,C = A\,(BC), \; \forall \; A, B, C \in M$, since multiplication of matrices is associative.
6. $A\,(B + C) = AB + AC$,

   and $(B + C)\,A = BA + CA \; \forall \; A, B, C \in M$, since matrix multiplication is distributive with respect to matrix addition.

Hence, M is a ring with respect to the two given compositions. The null matrix O of the type $n \times n$ is the zero element of this ring *i.e.*, $O - 0$.

Since the multiplication of matrices is not always commutative, therefore the ring is a non-commutative ring. [$n > 1$]

Now if I be the unit of the type $n \times n$, then $I \in M$ and we have $IA = A$ $AI \; \forall \; A \in M$. Therefore the matrix 1 is the multiplicative identity. Thus the ring is with unity and the matrix I is the unity element of the ring *i.e.*, $I = 1$.

1.5 SOME SPECIAL TYPES OF RINGS, RINGS WITH OR WITHOUT ZERO DIVISORS

Definition: *A non-zero element a of ring R is called a zero divisor or a divisor of zero if there exists an element $b \neq 0 \in R$ such that either $ab = 0$.*

Ring without Zero Divisors

A ring R is without zero divisors if the product of no two non-zero element of R is zero i.e., if

$$ab = 0 \Rightarrow a = 0 \text{ or } b = 0.$$

On the other hand we can say that, if in a ring R there exist non-zero elements a and b such that $ab = 0$, then R is said to be a ring with zero divisors.

Commutative Ring

A ring R is said to be commutative if the multiplication composition in R is commutative *i.e.*, of $a.b = b.a \; \forall \; a, b \in R$

Ring without Element

A ring R is said to be a ring with unity element if R has a multiplication identity, that is if a there exist an element in R denoted by 1 such that $1.a = a.1 = a \ \forall \ a \in R$

Example 1:

The ring of integers is a ring without zero divisors. The product of two non-zero elements is equal to the zero element of the ring.

Solution:

Do yourself.

Example 2:

The ring $(\{0, 1, 2, 3, 4, 5\}. +_6, \times_6)$ *is a ring with zero divisors. We have* $2 \times_6 3 = 0$, $3 \times_6 4 = 0$ *i.e., the product of two non-zero elements is equal to the zero element of the ring.*

Solution:

Do yourself.

Example 3:

Suppose M is a ring of all 2 × 2 matrices with their elements as integers, the addition and multiplication of matrices being the two ring compositions. Then M a ring with zero divisors.

The null matrix $O = \begin{bmatrix} 0 & 0 \\ 0 & 0 \end{bmatrix}$ is the zero element of this ring.

Now $A = \begin{bmatrix} 1 & 0 \\ 0 & 0 \end{bmatrix}$, $B = \begin{bmatrix} 0 & 0 \\ 1 & 0 \end{bmatrix}$ are two non-zero elements of this ring *i.e.*, $A \neq O$, $B \neq O$. We have

$$AB = \begin{bmatrix} 1 & 0 \\ 0 & 0 \end{bmatrix} \begin{bmatrix} 0 & 0 \\ 1 & 0 \end{bmatrix} = \begin{bmatrix} 0 & 0 \\ 0 & 0 \end{bmatrix} = O.$$

Thus the product of two non-zero elements of the ring is equal to the zero element of the ring. Therefore M is a ring with zero divisors.

Also it is interesting to note that

$$BA = \begin{bmatrix} 0 & 0 \\ 1 & 0 \end{bmatrix} \begin{bmatrix} 1 & 0 \\ 0 & 0 \end{bmatrix} = \begin{bmatrix} 0 & 0 \\ 1 & 0 \end{bmatrix} \neq \begin{bmatrix} 0 & 0 \\ 0 & 0 \end{bmatrix}.$$

Thus, in a ring R it is possible that ab = 0 but ba ≠ 0.

1.6 CANCELLATION LAWS IN A RING

Let R is a ring then R is an abelian group with respect to addition. For addition composition the cancellation laws hold in all rings. Therefore the question of cancellation laws holding in a ring arises only for the multiplication composition.

We say that cancellation laws hold in a ring R if

$$a \neq 0,\ ab = ac \Rightarrow b = 0$$

and $\quad a \neq 0,\ ab = ac \Rightarrow b = 0$ where a, b; c ∈ R.

Theorem:

A ring R is without zero divisors if and only if the cancellation laws hold in R i.e., R is without zero divisors ⇔ cancellation laws hold in R.

Proof:

Let R has no zero divisors. Let a, b, c be any three elements of R such that a ≠ 0, ab = ac.

We have ab = ac ⇒ ab − ac = 0 ⇒ a (b − c) = 0.

Since R is without zero divisors, therefore we have

$$a\,(b - c) = 0 \text{ and } a \neq 0 \Rightarrow b - c = 0 \text{ i.e., } b = c.$$

Thus the left cancellation law holds in R. Similarly, we can prove that the right cancellation law holds in R.

Conversely suppose that the cancellation laws hold in R. If possible let ab = 0, a ≠ 0, b ≠ 0.

Then we have ab = a0, since a0 = 0.

Now a ≠ 0, ab = a0 ⇒ b = 0 by left cancellation law.

Thus we get a contradiction. Hence R is without zero divisors.

1.7 INTEGRAL DOMAIN, FIELDS, DIVISION, RINGS, INTEGRAL, DOMAIN

Definition: *A ring is called an integral domain if it (1) is commutative, (2) has unit element, (3) is without zero divisors.*

A ring known as an integral domain if:

(i) It is commutative.

(ii) It has unit element.

(iii) It is without zero divisor. **Or**

An integral domain is a commutative ring with unity, which has no proper divisor or zero.

Example:

Ring I as integers is an example of integral domain. We have proved that I is a commutative ring with unity. Also I does not possess zero divisors. We know that if a, b are integers such that ab = 0, then either a or b must be zero.

Also it can be seen that the algebraic structures (C, +, .), (Q, +, .), (R, +, .) are all integral domains. As an example of a finite integral domain we have the ring ({0, 1, 2, 3, 4}, $+_5$, $\times_5$).

Inversible Elements in a Ring with Unity: *If R is a ring with unity, then a element a ∈ R is called inversible, if there exists b ∈ R such that ab = 1 = ba. Also then we write* $b = a^{-1}$.

Examples:

(i) 1 and –1 are the only two inversible elements of the ring of all integers.

(ii) n × n non-singular matrices with real numbers as elements are the only inversible elements of the ring of all n × n matrices with elements as real numbers.

1.8 DIVISION RING OR SKEW FIELD

Definition: *A ring R with at least two elements is called a division ring or a skew field if it (i) has unity, (ii) is such that each non-zero element possesses multiplicative inverse.*

Thus a commutative division ring is a field.

Every field is also a division ring. But a division ring is a field if it is also commutative. We shall later on give an example of a skew field which is not commutative *i.e.*, which is not a field.

1.9 FIELD

Definition: *A ring R with at least two elements is called a field if it:*

(i) is commutative,

(ii) has unity

(iii) is such that each non-zero element possesses multiplicative inverse.

Example:

The ring of rational numbers (Q, +, .) is a field since it is a commutative ring with unity and each non-zero element is inversible.

The rings of real numbers and complex numbers.

Theorem 1:

A finite commutative ring without zero divisors is a field.

Or

Every finite integral domain is a field

Proof:

Let us consider D be a finite commutative ring without zero divisors having n elements $a_1, a_2 ..., a_n$. In order to prove that D is a field, we must produce an element $1 \in D$ such that $1a = a \ \forall \ a \in D$. Also we should show that for every element $a \neq 0 \in D$ there exists an element $b \in D$ such that $ba = 1$.

Let $a \neq 0 \in D$. Consider the n products $aa_1, aa_2 ..., aa_n$.

All these are elements of D. Also they are distinct. For suppose that $aa_i = aa_j$ for $i \neq j$.

Then $\quad a(a_i - a_j) = 0.$ ...(i)

Since D is without zero divisors and $a \neq 0$, therefore (1) implies

$a_i - a_j = 0 \Rightarrow a_i = a_j$, contradicting $i \neq j$.

$\therefore$ $aa_1, aa_2 ..., aa_n$ are all the n distinct elements of D placed in some order. So one of these elements will be equal to a. Thus there exists an element, say, $1 \in D$ such that

$$a1 = a = 1a. \qquad [\because \text{ D is commutative}]$$

We shall show that this element 1 is the multiplicative identity of D. Let y be any arbitrary element of D. Then from the above discussion for some $x \in D$, we shall have $ax = y = xa$.

Now $\quad 1y = 1(ax) \qquad [\because ax - y]$

$= (1a)x$

$= ax \qquad [\because 1a = a]$

$= y \qquad [\because ax = y]$

$= y1$ [$\because$ D is commutative]

Thus, $ly = y = y1$, $\forall\ y \in D$. Therefore 1 is the unit element *i.e.*, the multiplicative identity of the ring D.

Now $1 \in D$. Therefore from the above discussion one of the n products $aa_1, aa_2 ..., aa_n$ will be equal to 1. Thus there exists an element, say, $b \in D$ such that

$$ab = 1 = ba.$$

$\therefore$ b is the multiplicative inverse of the non-zero element $a \in D$. Thus every non-zero element of D is inversible.

Hence D is a field. **Hence Proved.**

Theorem 2:

Every field is an integral domain.

Proof:

Let a, b be elements of F with $a \neq 0$ such that $ab = 0$.

Since $a \neq 0$, a^{-1} exists and we have

$$ab = 0 \Rightarrow a^{-1}(ab) = a^{-1}\,0$$

$$\Rightarrow (a^{-1}a)b) = 0$$

$$\Rightarrow 1b = 0 \qquad [\because a^{-1}a = 1]$$

$$\Rightarrow b = 0. \qquad [\because 1b = b]$$

Similarly, let ab = and $b \neq 0$.

Since $b \neq 0$, b^{-1} exists and we have

$$ab = 0 \Rightarrow (ab)\,b^{-1} = 0\,b^{-1}$$

$$\Rightarrow a\,(bb^{-1}) = 0 \Rightarrow al = 0 \Rightarrow a = 0.$$

Thus in a field $ab = 0 \Rightarrow a = 0 \Rightarrow b = 0$. Therefore a field has no zero divisors. Therefore every field is an integral domain.

But the converse is a not true i.e., every integral domain is not a field. For example the ring of integers is an integral domain and it is not a field. The only inversible elements of the ring of integers are 1 and –1.

Note: A field has no zero divisors. Therefore in a field the product of two non-zero elements will again be a non-zero element. Also the unit element $1 \neq 0$ and each non-zero element possesses multiplicative inverse which is again a non-zero element. The multiplication is commutative as well as associative. Therefore *the non-zero elements of a field form an abelian group with respect to multiplication.*

Theorem 3:

As field (skew-field) has no divisors of zero.

Proof:

Let D be a skew-field. Then D is a ring with unit element 1 and each non-zero element of D possesses multiplicative inverse.

Let a, b be elements of D with $a \neq 0$ such that $ab = 0$.

Since $a \neq 0$, a^{-1} exists and we have

$$ab = 0 \Rightarrow a^{-1}(ab) = a^{-1} 0$$

$$\Rightarrow (a^{-1} a) b = 0 \Rightarrow 1b = 0 \Rightarrow b = 0.$$

Similarly, let $ab = 0$ with $b \neq 0$.

Since $b \neq 0$, b^{-1} exists and we have

$$ab = 0 \Rightarrow (ab) b^{-1}$$

$$\Rightarrow a (bb^{-1}) = 0 \Rightarrow a1 = 0 \Rightarrow a = 0.$$

Therefore a skew-field has no zero divisors. **Hence Proved.**

Example 1:

Show that $Z(\sqrt{-5})$, the set of complex numbers $a + b\sqrt{(-5)}$ where a, b are integers, is an integral domain.

Solution:

Let $a + b\sqrt{(-5)}$ and $c + d\sqrt{(-5)}$ be any two complex numbers belonging to the set $Z[\sqrt{-5}]$, where a, b, c, d are any integers.

Then $[a + b\sqrt{(-5)}] + [c + d\sqrt{(-5)}] = (a + c) + (b + d)\sqrt{(-5)}$

and $[a + b\sqrt{(-5)}][c + d\sqrt{(-5)}] = (ac - 5bd) + (ad + bc)\sqrt{(-5)}$.

There are again members of the set $Z[\sqrt{-5}]$ because $a + c$, $b + d$, $ac - 5bd$, $ad + bc$ are all integers. Therefore $Z[\sqrt{-5}]$ is closed with respect ordinary addition and multiplication of complex numbers.

Further in complex numbers both addition and multiplication are associative as well as commutative compositions. Also multiplication distributes with respect to addition. The complex number $0 + 0\sqrt{(-5)}$ is a member of the set $Z[\sqrt{-5}]$ and is the additive identity. The additive inverse of $a + b\sqrt{(-5)} \in Z[\sqrt{-5}]$ is $(-a) + (-b)\sqrt{(-5)}$. The complex number $1 + 0\sqrt{(-5)}$ is a member of the set $Z[\sqrt{-5}]$ and is the multiplicative identity.

Therefore the set of complex numbers $a + b\sqrt{(-5)}$ where a, b, are integers is a commutative ring with unity for the addition and multiplication of complex numbers as the two ring compositions.

Also this ring is free from zero divisors since the product of two non-zero complex numbers cannot be zero. Therefore Z $[\sqrt{-5}]$ is an integral domain for ordinary addition and multiplication of complex numbers as the two ring compositions.

Example 2:

Do the following sets form integral domains with respect to ordinary addition and multiplication? If so state if they are fields.

1. *The set of numbers of the form b $\sqrt{2}$ with b rational.*
2. *The set of even integers.*
3. *The set of positive integers.*

Solution:

1. Let A = $\{b\sqrt{2} : b \in Q\}$.

   We have $3\sqrt{2} \in A$ and $5\sqrt{2} \in A$. Then $(3\sqrt{2})(5\sqrt{2}) = 30$.

   Now 30 cannot be put in the form b $\sqrt{2}$ where b is a rational number. Therefore $30 \notin A$. Thus A is not closed with respect to multiplication. Therefore the question of A becoming a ring does not arise.
2. Let R be the set of all even integers. Then R is a ring with respect to addition and multiplication of integers. Also the multiplication is a commutative composition. r is without zero divisors since the product of two non-zero even integers cannot be equal to zero which is the zero element of this ring. Since the integer $1 \notin R$, therefore R is a ring without unity.

   R will be an integral domain if we do not require the existence of the unit element for an integral domain.

   But R is not a field since the multiplicative identity does not exist.
3. Let N be the set of positive integers. Since the integer $0 \notin N$, therefore the additive identity does not exist. So N will not be a ring.

Example 3:

An element a of a ring R is said to be nilpotent if $a^n = 0$ for some positive integer n. Prove that $a = 0$ is the only nilpotent element of an integral domain.

Solution:

Let R be an integral domain. We have $0^n = 0$ for every positive integer n.

$\therefore$ The element 0 of R is surely nilpotent.

Now let a be any non-zero element of R. Since in an integral domain the product of two non-zero elements cannot be zero, therefore

$$a \neq 0 \Rightarrow a^2 = aa \neq 0 \Rightarrow a^3 = aa^2 \neq 0 \Rightarrow a^4 = aa^3 \neq 0, \text{ and so on.}$$

Thus, in an integral domain if $a \neq 0$, then $a^n \neq 0$ for every positive integer n and so a cannot be nilpotent.

Hence 0 is the only nilpotent element of an integral domain.

Example 4:

Prove that the set I (√2) of number of the form a + b √2, with a and b as integers is an integral domain with respect to ordinary addition and multiplication. Is it a field ?

Solution:

We can easily verify that the given system is a commutative ring with unity element $1 + 0\sqrt{2}$. Also $0 + 0\sqrt{2}$ is the zero element of this ring. Now in order to prove that its ring is an integral domain, we should prove that this ring is with out zero divisors.

Let $a + b\sqrt{2}$ and $c + d\sqrt{2}$ be any two elements of this ring. Then

$$(a + b\sqrt{2})(c + d\sqrt{2}) = 0 + 0\sqrt{2}$$

$\Rightarrow$ ac + 2bd = 0 and bc + ad = 0 and this will happen only when either

$$a = 0 \text{ and } b = 0$$

$$\Rightarrow \quad c = 0 \text{ and } d = 0.$$

Thus $(a + b\sqrt{2})(c + d\sqrt{2}) = 0 + 0\sqrt{2}$

$$\Rightarrow \text{either } a + b\sqrt{2} = 0 \Rightarrow e + d\sqrt{2}.$$

Thus, the given ring is without zero divisors. Therefore it is an integral domain. But it is not a field. Obviously $5 + 3\sqrt{2}$ is a non-zero element of this ring. Its inverse would have been

$$\frac{1}{5+3\sqrt{2}} = \frac{5-3\sqrt{2}}{\left(5+3\sqrt{2}\right)\left(5-3\sqrt{2}\right)} = \frac{5-3\sqrt{2}}{7} = \frac{5}{7} - \frac{3}{7}\sqrt{2}$$

which is not an element of this ring.

Example 5:

Let p be a prime number. Prove that the set of integers I_p,

$$I_p = \{0, 1, 2, 3, ..., p-1\}$$

Forms a field with respect o addition and multiplication modulo p.

Solution:

As in the chapter on groups we can prove that $(I_p, +_p)$ is an abelian group. Also I_p is closed with respect to '$\times_p$'. The composition '$\times_p$' is distributive with respect to '$+_p$' as can be proved as below:

Let a, b, c ∈ I. Then

$a \times_p (b +_p c) = a \times_p (b + c)$ $[\because\ b =_r c \equiv b + c \pmod p]$

= least non- negative remainder when a (b + c) is divided by p

= least non- negative remainder when ab + ac is divided by p

$= (ab) +_p (ac)$

$= (a \times_p b) +_p (ac)$ $[\because\ ab \equiv b \times_r b \pmod p]$

$= (a \times_p b) +_p (a \times_p c).$

Similarly we can prove the other distributive law.

1 is the identity element for ' $\times_p$'.

The zero element of the ring $(I_p\ +_p\ \times_p)$ is 0. As in groups we can prove that each non-zero of I_p has multiplicative inverse.

In short the non-zero elements of I_p from an abelian group with respect to '$\times_p$'. Therefore $(I_p, +_p, \times_p)$ is a field.

Note: If p is not prime, then this ring will have zero divisors and so it cannot be a field.

Example 6:

Prove that the only idempotent elements of an integral domain with unity are 0 and 1.

Solution:

Let R be an integral domain with unity 1.

Let a ∈ R and a be idempotent *i.e.*, $a^2 = a$.

We have $a^2 = a \Rightarrow aa = a1$

$\Rightarrow aa - a1 = 0 \Rightarrow a(a - 1) = 0.$

But in an integral domain the product of two elements is 0 only if at least one of them is zero.

$\therefore \quad a(a - 1) = 0 \Rightarrow a = 0 \Rightarrow a - 1 = 0$

$\Rightarrow a = 0 \Rightarrow a = 1.$

Hence the only idempotent elements of an integral domain with unity are 0 and 1.

Example 7:

A Gaussian integer is a complex number a + ib, where a and b are integers. Show that the set J[i] of Gaussian integers forms a ring under ordinary addition and multiplication of complex numbers. Is it an integral domain? Is it a field?

Solution:

Let a + ib and c + id be any two Gaussian integers.

Then $(a + ib) + (c + id) = (a + c) + i(b + d)$

and $(a + ib) + (c + id) = (ac - bd) + i(ad + bc)$.

These are again Gaussian integers. Therefore J[i] is closed with respect to ordinary addition and multiplication of complex numbers.

Further in complex numbers both addition and multiplication are associative as well as commutative compositions. Also multiplication distributes with respect to addition. The Gaussian integer 0 + i0 is the additive identity. The additive inverse of a + ib is (–a) + i (–b). The Gaussian integer 1 + i0 is the multiplicative identity.

Therefore the set of Gaussian integers is a commutative ring with unity for the given compositions.

Also this ring is free from zero divisors since the product of two non-zero complex numbers cannot be zero. Therefore J[i] is an integral domain.

But this is not field since the multiplicative inverse of a + ib will be $\frac{a}{a^2 + b^2} + i\left(-\frac{b}{a^2 + b^2}\right)$ which is not always a Gaussian integer as $\frac{a}{a^2 + b^2}$ and $-\frac{b}{a^2 + b^2}$ are not necessarily integers

Example 8:

If two operations $*$ *and* o on the set I of integers are defined as follows.

$$a * b = a + b - 1,\ a \circ b = a + b - ab,$$

prove that the system $(I, *, \circ)$ *is a commutative ring with identity.*

Solution:

First we shall show that the algebraic structure $(I, *)$ is an abelian group.

In is closed for the operation $*$. Let a, b, ∈ I.

Then $a * b = a + b - 1$ which is also a member of I. Therefore I is closed with respect to the operation *.

Commutativity and associativity of the operation * on these I.

Let a, b, c $\in$ I. Then

$$a * b = a + b - 1 = b + a - 1 = b * a$$

and $(a * b) * c = (a + b - 1) * c = (a + b - 1) + c - 1 = a + (b + c - 1) - 1$

$$= a * (b + c - 1) = a * (b * c).$$

Thus the operation * on the set I is commutative as well as associative.

Existence of Identity for the Operation * on I

The integer e $\in$ I will be the identity for the operation * on the set I if

$e * a = a \ \forall \ a \in I.$ [Note that $e * a = a * e$]

Now $e * a = a \Rightarrow e + a - 1 = a \Rightarrow 0 \Rightarrow e = 1.$

Therefore the integer 1 is identity for the operation * on the set I and will be the zero element of the ring c I, *, o).

Existence of Inverse of Each Element of I for the Operation *

Let a $\in$ I. Then b $\in$ I will be the inverse of a for the operation * if $a * b = 1.$

Now $a * b = 1 \Rightarrow a + b - 1 = 1 \Rightarrow b = 2 - a.$

Thus if a $\in$ I, then $2 - a \in I$ and is the inverse of a for the operation *.

$\therefore$ (I, *) is an abelian group.

I is closed for the operation o. Let a, b $\in$ I.

Then $a \circ b = a + b - ab$ which is also an integer. Therefore I is closed with respect to the operation o.]

Commutativity and associativity of the operation o on the set I. Let a, b, c $\in$ I. Then

$$a \circ b = a + b - ab = b + a - ba = b \circ a$$

Also $(a \circ b) \circ c = (a + b - ab) \circ c = (a + b - ab) + c - (a + b - ab) c$

$$= a + b + c - ab - ac - bc + abc$$

and $a \circ (b \circ c) = a \circ (b + c - bc) = a + (b + c - bc) - a (b + c - bc)$

$$= a + b + c - bc - ab - ac + abc.$$

$\therefore a \circ (b \circ c) = (a \circ b) \circ c.$

Thus, the operation o on the set I is commutative as well as associative.

Distributivity of o Over $*$

Let a, b, c $\in$ I. Then

$a \circ (b * c) = a \circ (b + c - 1) = a\,(b + c - 1) - a\,(b + c - 1)$

$= a + b\,c - 1 - ab - ac + a = 2a + b + c - ab - ac - 1.$

Also $(a \circ b) * (a \circ b) = (a + b - ab) * (a + c - ca)$

$= a + b - ab + a + c - ac - 1 = 2a + b + c - ab - ac - 1.$

$\therefore a \circ (b * c) = (a \circ b) * (a \circ c).$

Similarly we can show that the other distributive law

$(b * c) \circ a = (b \circ a) * (c \circ a)$

also holds good.

Existence of Identity for the Operation o on I

The integer u $\in$ I will be the identity for the operation o on the set I if

$u \circ a = a \;\forall\; a \in I.$ [Note $u \circ a = a \circ u$]

Now $u \circ a = a \;\forall\; a \in I$

$\Rightarrow\; u + a - ua = a \;\forall\; a \in I$

$\Rightarrow\; u\,(1 - a) = 0 \;\forall\; a \in I$

$\Rightarrow\; u = 0.$

Therefore the integer 0 is the identity for the operation o on the set I and so will be the unity element of this ring with unity. We observe that $\forall\; a \in I$, we have

$0 \circ a = 0 + a - 0a = a = a \circ 0.$

Hence the algebraic structure (I, $*$ o) is a commutative ring with identity. The operation $*$ is the addition of this ring and the operation o is the multiplication of this ring. The integer 1 is the zero element of this ring and the integer 0 is the unity element of this ring with unity.

Example 9:

If a, b are any elements of a ring R, prove that

1. $-(-a) = a;$
2. $-(a + b) = -a - b;$
3. $-(a - b) = -a + b.$

Solution:

1. Since R is a group with respect to addition, therefore $-(-a) = a$. [Remember that in a group $(a^{-1})^{-1} = a$].
2. Since R is a group with respect to addition, therefore $-(a + b) = (-b) + (-a)$. But addition in R is commutative.

$\therefore \quad (-b) + (-a) = (-a) + (-b) = -a - b.$

$\therefore \quad -(a + b) = -a - b.$

3. $-(a - b) = -[a + (-b)] = -a + [-(-b)] = -a + b.$

Example 10:

Prove that if a, b ∈ R then $(a + b)^2 = a^2 + ab + ba + b^2$, *where by* x^2 *we mean xx.*

Solution:

We have

$(a + b)^2 = (a + b)(a + b)$

$= a(a + b) + b(a + b)$ [by right distributive law]

$= (aa = ab) + (ba + bb).$ [by left distributive law]

$= a^2 = ab = ba + b^2.$

Example 11:

If a, b, c are elements of a ring R, then evaluate (a + b) (c + d).

Solution:

We have

$(a + b) + (c + d) = a(c + d) + b(c + d)$ [by right distributive law]

$= ac + ad + bc = bd.$ [by left distributive law]

Example 12:

Prove that the set of residue classes modulo p is a commutative ring with respect to addition and multiplication of residue classes. Further show that the ring of residue classes modulo p is a field if and only if p is a prime.

Solution:

Let I_p be the set of residue classes modulo p. Then I_p has p distinct elements. Thus $I_p = \{[0], [1], [2], ..., [p - 1]\}$.

Let $[a], [b] \in I_p$. Then we define addition and multiplication or residue classes as follows:

$$[a] + [b] = [a + b], \quad \text{(addition)}$$

$$[a]\,[b] = [ab] \quad \text{(multiplication)}$$

Since $[a + b]$ and $[ab]$ are both residue classes modulo p, therefore I_p is closed with respect to addition and multiplication.

Now let $[a], [b], [c]$ be any elements of I_p. Then we observe:

Commutativity of Addition:

$$[a] + [b] = [a + b] \quad \text{(by def, of addition of residue classes)}$$

$$= [b + a] \quad (\because \text{ addition of integers is commutative})$$

$$= [b] + [a].$$

Associativity of Addition: We have

$$([a] + [b]) + [c] = [a + b] + [c] = [(a + b) + c] = [a + (b + c)]$$

$$= [a] + [b + c] = [a] + ([b] + [c]).$$

Existence of Additive Inverse: We have $[0] \in I_p$. If $[a] \in I_p$, then $[0] + [a] = [0 + a] = [a]$. Therefore $[0]$ is the additive identity.

Existence of Additive Inverse: Let $[a] \in I_p$. Then $[-a] \in I_p$. We have $[-a] + [a] + [a] = [-a + a] = [0]$. Therefore $[-a]$ is the additive inverse of $[a]$.

Associativity of Multiplication: We have

$$([a]\,[b])\,[c] = [ab]\,[c] = [(ab)\,c] = [a\,(bc)] = [a]\,[bc] = [a]\,([b]\,[c]).$$

Commutativity of Multiplication: We have

$$[a]\,[b] = [ab] = [ba] = [b]\,[a].$$

Distributive Laws: We have

$$[a]\,([b] + [c]) = [a]\,[b + c] = [a\,(b + c)] = [ab + ac]$$

$$= [ab] + [ac] = [a]\,[b] + [a]\,[c].$$

Similarly $([b] + [c])\,[a] = [b]\,[a] + [c]\,[a]$.

Thus, I_p is a commutative ring. Note that it is a finite ring because it has p elements. Now suppose that p is a prime number. Then to prove that I_p is a field. Let $[a], [b] \in I_p$. Then

$[a]\,[b] = [0]$

$\Rightarrow [ab] = [0]$

$\Rightarrow$ p is a divisor of ab *i.e.*, $p \mid ab$

$\Rightarrow p \mid a$ or $p \mid b$. [Note that if a and b are any two integers and p is a prime number, then $p \mid ab \Rightarrow p \mid a \Rightarrow p \mid b$]

$\Rightarrow [a] = [0] \Rightarrow [b] = [0]$.

Thus, I_p is without zero divisors. Therefore I_p is an integral domain. But every finite integral domain is a field. Hence I_p is a field. Conversely suppose that I_p is a field. Then I_p is an integral domain. Therefore I_p is without zero divisors. We are to prove that p is a prime number. Suppose p is not prime, but p is composite. Let $p = mn$, where $1 < m < p$, $1 < n < p$. Then

$$[mn] = [p]$$

$$\Rightarrow [m]\,[n] = [0]. \qquad (\because\ [p] = [0])$$

Also $[m] \neq [0]$, since $1 < m < p$. Similarly $[n] \neq 0$.

Thus $[m]\,[n] = [0]$ while neither $[m] = [0]$ nor $[n] = [0]$.

Therefore I_p possesses zero divisors and thus we get a contradiction. Hence p must be prime.

Example 13:

If a, b, c, d are any elements of a ring R, prove that

1. $(a - b)(c - d) = (ac + bd) - (ad + bc)$.
2. $a - b = c - d \Leftrightarrow a + d = b + c$.

Solution:

1. We have

$$(a - b)(c - d) = (a - b)c - (c - b)d \qquad [\because\ a(b - c) = ab - ac]$$
$$= (ac - bc) - (ad - bd) = (ac - bc) - ad + bd$$
$$= (ac + bd) - bc - ad$$
$$[\because \text{ addition is commutative and associative}]$$
$$= (ac + bd) - (bc + ad).$$

2. We have $a - b = c - d$

$\Leftrightarrow a + (-b) = c + (-d)$

$\Leftrightarrow a + (-b) + d = c + (-d) + d$

$\Leftrightarrow a + d + (-b) = c + 0$

[$\because$ in a ring addition is commutative as well as associative and $-d = d = 0$]

$\Leftrightarrow a + d + (-b) = c \qquad [\because\ c + 0 = c]$

$\Leftrightarrow a + d + (-b) + d = c + (-d) + d$

$\Leftrightarrow a + d + 0 = c + b \Leftrightarrow a + d = c + b.$

Example 14:

Prove that in a field

1. $\frac{a}{b} = \frac{c}{d} \Leftrightarrow ad = bc.$

2. $\frac{a}{b} - \frac{c}{d} = \frac{ad - bc}{bd}$

3. $(-a)^{-1} = -(a^{-1}).$

4. $\frac{(-a)}{(-b)} = \frac{a}{b}.$

Solution:

Let a, b, c, d $\in$ F where F is a field. Here $b \neq 0$, $d \neq 0$.

1. We have $\frac{a}{b} = \frac{c}{d} \Leftrightarrow ab^{-1} = cd^{-1} \Leftrightarrow (ab^{-1})\ (bd = (cd^{-1})\ (bd)$

$\Leftrightarrow (ad)\ (b^{-1}\ b) = (bc)\ (d^{-1}\ d)$ [$\because$ in a field multiplication is commutative as well as associative]

$\Leftrightarrow (ad)\ 1 = (bc)\ 1 \Leftrightarrow ad = bc.$

2. We have $\frac{a}{b} - \frac{c}{d} = (ab^{-1}) - (cd^{-1})$

$= (bd)^{-1}\ (bd)\ [(ab^{-1}) - (cd^{-1})]$

$= (bd)^{-1}\ [(bd)\ (ab^{-1}) \quad (bd)\ (cd^{-1})] = (bd)^{-1}\ (ad - bc) = \frac{ad - bc}{bd}.$

3. We have $(-a)\ [-(a^{-1})] = aa^{-1}$ [$\because$ in a field $(-a)\ (-b) = ab$] $= 1.$

$\therefore \quad (-a)^{-1} = -(a^{-1}).$ [$\because$ in a field $ab = 1 \Leftrightarrow a^{-1} = b$]

4. We have $\frac{(-a)}{(-b)} = (-a)\ (-b)^{-1} = (-a)\ [-(b^{-1})]$

[$\because$ by part (3) of this question $(-b)^{-1} = -(b^{-1})$]

$= ab^{-1}$ [$\because$ in a field $(-a)\ (-b) = ab$]

$= \frac{a}{b}.$

Example 15:

Prove that the set M of 2 × 2 matrices over the field of real numbers is a ring with respect to matrix addition and multiplication. It is a commutative ring with unity element ? Find the zero element. Does this ring possess zero divisors?

Solution:

Let A, B, ∈ M. Then A + B ∈ M and AB ∈ M. Therefore M is closed with respect to addition and multiplication of matrices.

$$\therefore \quad A + (B + C) = (A + B) + C \ \forall \ A, B, C \in M$$

and
$$A (BC) = (AB) C \ \forall \ A, B, C \in M.$$

Addition of matrices is a commutative composition. Therefore for all A, B ∈ M, we have A + B = B + A.

If O be the null matrix of the type 2 × 2, then O ∈ M and O + A + A ∀ A ∈ M.

Further multiplication of matrices is distributive with respect to addition.

$$\therefore \quad A (B + C) = AB + AC$$

and
$$(B + C) A = BA + CA \ \forall \ A, B, C \in M.$$

∴ M is a ring with respect to the given compositions.

Multiplication of matrices is not in general a commutative composition.

For example, if $A = \begin{bmatrix} 2 & 4 \\ 3 & 5 \end{bmatrix}$, $B = \begin{bmatrix} 1 & 2 \\ 0 & 1 \end{bmatrix}$, then

$$AB = \begin{bmatrix} 2 & 4 \\ 3 & 5 \end{bmatrix} \begin{bmatrix} 1 & 2 \\ 0 & 1 \end{bmatrix} = \begin{bmatrix} 2 & 8 \\ 3 & 11 \end{bmatrix} \text{ and } BA = \begin{bmatrix} 1 & 2 \\ 0 & 1 \end{bmatrix} \begin{bmatrix} 2 & 4 \\ 3 & 5 \end{bmatrix} = \begin{bmatrix} 8 & 14 \\ 3 & 5 \end{bmatrix}$$

Thus AB ≠ BA and so the ring is a non-commutative ring.

If I be the unite matrix of the type 2 × 2 *i.e.*, if $I = \begin{bmatrix} 1 & 0 \\ 0 & 1 \end{bmatrix}$, then I ∈ M. Also we have AI = A = IA ∀ A ∈ M.

∴ I is the multiplicative identity.

Thus the ring possesses the unit element and we have I = 1 (the unit element of the ring).

Thus mull matrix $O = \begin{bmatrix} 0 & 0 \\ 0 & 0 \end{bmatrix}$ is the additive identity and is therefore the zero element of the ring *i.e.*, O = 0 (the zero element of the ring).

The ring possesses zero divisors. For example if

$$A = \begin{bmatrix} 0 & 1 \\ 0 & 1 \end{bmatrix}, B = \begin{bmatrix} 2 & 3 \\ 0 & 0 \end{bmatrix}, \text{ then } AB = \begin{bmatrix} 0 & 1 \\ 0 & 1 \end{bmatrix}\begin{bmatrix} 2 & 3 \\ 0 & 0 \end{bmatrix} = \begin{bmatrix} 0 & 0 \\ 0 & 0 \end{bmatrix}$$

Thus, the product of the two non-zero elements of the ring is equal to the zero element of the ring.

Example 16:

If R is a system satisfying all the conditions for a ring with unit element with the possible exception a+b=b+a, prove that the axiom a +b = b + a must hold in R and that R is thus a ring.

Solution:

Since 1 is an element of R, we have

$(a + b)(1 + 1) = a(1 + 1) + b(1 + 1)$ [by right distributive law]

$= (a1 + a1) + (b1 + b1) = (a + a) + (b + b)$ ...(i)

Also $(a + b)(1 + 1) = (a + b)1 + (a + b)1$ [by left distributive law]

$= (a + b) + (a + b)$ ...(ii)

[$\because$ 1 is the unit element]

From (i) and (ii), we get

$(a + a) + (b + b) = (a + b) + (a + b)$

$\Rightarrow [(a + a) + b] + b = [(a + b) + a] + b$ [by associativity of addition]

$\Rightarrow (a + a) + b = (a + b) + a$ [by right cancellation law for addition in R since with the given postulates R is a group with respect to addition].

$\Rightarrow a + (a + b) = a + (b + a)$ [by associativity of addition in R]

$\Rightarrow (a + b) = (b + a)$ [by left cancellation law for addition in R]

Thus addition is commutative in R. Hence R is a ring.

Example 17:

Prove that in the list of axioms for a ring R with unity the axiom demanding commutativity under addition may be omitted.

Solution:

Suppose an algebraic structure (R, + , .) satisfies all the axioms for a ring with unity except the axiom of commutativity of addition. Then we have to show that addition must be commutative on R and thus R must be a ring.

Example 18:

If R is a commutative ring, prove by induction that

$$(a + b)^n = a^n + {}^nC_1 a^{n-1}b + {}^nC_2 a^{n-2}b^2 + \ldots + b^n$$

for every positive integer n; here a and b are elements of R.

Solution:

We have $a + b = a^1 + b^1$.

Now $(a + b)^2 = (a + b)(a + b)$

$= a^2 + ab + ba + b^2$, by list. Laws

$= a^2 + ab + ab + b^2$ [$\because$ ab = ba, the ring R being commutative]

$= a^2 + 2ab + b^2$

$= a^2 + {}^2C_1 ab + b^2$.

Thus the theorem is true for n = 2.

Now assume that the theorem is true for any positive integer n *i.e.*,

$$(a + b)^n = a^n + {}^nC_1\, a^{n-1}b + \ldots + {}^nC_r\, a^{n-1}b^r + {}^nC_{r+1}\, a^{n+r-1}\, b^{r+1} + \ldots + b^n.$$

Then $(a + b)^{n+1} = (a + b)(a + b)^n$

$= (a + b)(a^n + {}^nC_1\, a^{n-1}b + {}^nC_2\, a^{n-2}b^2 + \ldots$
$+ {}^nC_r\, a^{n-1}b^r + {}^nC_{r+1}\, a^{n-r-1}\, b^{r+1} + \ldots + b^n)$

$= a^{n+1} + (ba^n + {}^nC_1 a^n b) + \ldots$
$+ ({}^nC_r\, ba^{n-r}\, b^r + {}^nC_{r+1}\, a^{n-r}\, b^{r+1}) + \ldots + b^{n+1}$.

Since the ring R is commutative, therefore

$ab^n = a^n b$, $ba^{n-r} = (ba^{n-r})\, b^r = (a^{n-r}\, b)\, b^r$

$= a^{n-r}\, bb^r = a^{n-r}\, b^{r+1}$.

Also $1 = {}^nC_0$, ${}^nC_r + {}^nC_{r+1} = {}^{n+1}C_{r+1}$.

Hence $(a + b)^{n+1} = a^{n+1} + ({}^nC_0 + {}^nC_1)\, a^n b + \ldots$
$+ ({}^nC_r + {}^nC_{r+1})\, a^{n-r}\, b^{r+1} + \ldots + b^{n+1}$

$= a^{n+1} + {}^{n+1}C_1\, a^{(n+1)-1}\, b + \ldots$
$+ {}^{n+1}C_{r+1}\, a^{(n+1)-(r+1)}\, b^{r+1} + \ldots + b^{n+1}$.

Thus, the theorem is true for n + 1, if it is true for n. But it is true for n = 1 and n = 2. Hence by induction it is true for all positive integral values of n.

Example 19:

If a ring R has a left identity as well as right identity, then the two are equal.

Solution:

Let R be a ring which possesses left identity e and right identity e' *i.e.*, $ea = a \ \forall \ a \in R$ and $ae' = a \ \forall \ a \in R$.

To prove that $e = e'$.

Since e is left identity and $e \in R$, therefore

$$ee' = e'. \quad (1)$$

Again e' is right identity and $e \in R$.

$$\therefore \quad ee' = e. \quad (2)$$

Since ee' is a unique element of R, therefore from (1) and (2) we can lead that eqn.

Example 20:

Prove that the set {0, 1, 2} (mod 3) is a field with respect to addition and multiplication.

Solution:

Let R = {0, 1, 2,}. To prove that $(R, +_3, \times_3)$ is a field. First we shall show that $(R, +_3)$ is an abelian group. The composition table for R for the operation $+_3$ is as given below.

$+_3$	0	1	2
0	0	1	2
1	1	2	0
2	2	0	1

From the composition table we see that R is closed for $+_3$ and the operation $+_3$ on the set R is commutative. The operation $+_3$ on the set R is also associative as we know it.

The element $0 \in R$ is identity for $+_3$. The inverses of 0, 1, 2, for $+_3$ are 0, 2, 1 respectively. Therefore $(R, +_3)$ is an abelian group.

Now we prepare composition table for R for the operation $\times_3$.

$+_3$	0	1	2
0	0	0	0
1	0	1	2
2	0	2	1

From the composition table we see that R is closed for $\times_3$ and the operation $\times_3$ on the set R is commutative.

The operation $\times_3$ on the set R is associative. For if a, b, c $\in$ R, then

$$a \times_3 (b \times_3 c) = (a \times_3 b) \times_3 c \text{ because } a (bc) = ab) c.$$

Also the operation $\times_3$ distributes over the operation $+_3$ For if a, b, c, $\in$ R, then

$$a \times_3 (b +_3 c) = (a \times_3 b) +_3 (a \times_3 c)$$

and $$(b +_3 c) \times_3 a = (b \times_3 a) +_3 (c \times_3 a).$$

The element $1 \in$ R is identity for the operation $\times_3$ because from the composition table we see that

$$1 \times_3 a = a \; a \times_3 1, \; \forall \; a \in R.$$

Also each non-zero element of R possesses multiplicative inverse. From the composition table we see that the inverses of the non-zero elements 1 and 2 of R for $\times_3$ are 1 and 2 respectively.

Thus, (R, $+_3$, $\times_3$) is a commutative ring with unity element 1 and each non-zero element of R is inversible. Hence (R, $+_3$, $\times_3$) is a field.

Example 21:

Is every field also a division ring ? Does the set of all integers under usual addition and multiplication form a field ? Give some example of a field which is finite.

Solution:

First give the definitions of a field and a division ring. For these definitions refer.

From these two definitions we conclude that every field is also a division ring.

The set of integers I under usual addition and multiplication is a ring. The ring of integers (I, +, .) is a commutative ring with unity element the integer 1. But this ring is not a field. In this ring the only inversible elements are 1 and –1 while in a field very non-zero element must be inversible.

Example of a Finite Field: The ring ({0, 1, 2,}, $+_3$, $\times_3$) is an example of a finite field because the number of distinct elements in this ring is 3 and so it is a finite ring. This ring is a commutative ring with unity element the integer 1. The zero elements 1 and 2 of this ring possess multiplicative inverse. The multiplicative inverse of 1 is 1 because $1 \times_3 1 = 1$ and the multiplicative inverse of 2 is 2 because $2 \times_3 2 = 1$.

Hence ({0, 1, 2,}, $+_3$, $\times_3$) is a field and it is a finite field.

Example 22:

Define a field. Prove that every field is an integral domain, but there exist some integral domains which are not fields.

Solution:

For the definition of a field refer to prove that every field is an integral domain refer

For an example of a ring which is an integral domain but is not a field, consider the ring of integers (I, +, .) where I is the set of integers *i.e.*,

$$I = \{..., -3, -2, -1, 1, 0, 2, 3, ...\}.$$

The ring of integers (I, +, .) is an integral domain because it is a commutative ring with unity 1 and does not possess zero divisors. But this ring is not a field.

The only elements of the ring of integers which possess multiplicative inverse are 1 and –1 while in a field every non-zero element must possess multiplicative inverse.

Hence, the ring of integers is an integral domain but it is not a field.

Example 23:

Prove that the set of integers R,

$$R = \{0, 1, 2, 3, 4\}$$

forms a field under addition modulo 5 and multiplication modulo 5.

Solution:

Do yourself.

Hint : Hence also 0 is identity for the operation $+_5$. Also for the operation $+_5$, the inverses of 0, 1, 2, 3, 4 are 0, 4, 3, 2, 1 respectively.

1 is multiplicative identity *i.e.*, identity for the operation $\times_5$. Also for the operation $\times_5$ the inverses of the non-zero elements 1, 2, 3, 4, of R are 1, 3, 2, 3 respectively.

Example 24:

Define a ring and an integral domain. Give an example of a ring which is not an integral domain.

Solution:

For the definition of a ring refer § 1 and for the definition of an integral domain refer § 6.

For an example of ring which is not an integral domain consider the ring $(R, +_6, \times_6)$ where $R = \{0, 1, 2, 3, 4, 5\}$.

The above ring $(R, +_6, \times_6)$ is a commutative ring and possesses unity element which is the integer 1. But this ring is not an integral domain because it possesses zero divisors.

We observe that $2 \in R$, $3 \in R$, $2 \neq 0$, $3 \neq 0$, but $2 \times_6 3 = 0$. Thus both 2 and 3 are zero divisors. Hence the ring

$$(\{0, 1, 2, 3, 4, 5\}, +_6, \times_6)$$

is not an integral domain.

Example 25:

Define a ring and furnish an example of (1) a non-commutative ring with unity, (2) a commutative ring without unity.

Solution:

For the definition a ring refer § 1.

1. *Example of a non-commutative ring with unity:* Let M be the set of all 2×2 matrices with elements as real numbers. Then M is a ring for addition and multiplication of matrices as the two ring operations.

This ring is a non-commutative ring because the operation of multiplication of matrices on the set M is not commutative.

For example, if we take $A = \begin{bmatrix} 1 & 2 \\ 3 & 5 \end{bmatrix}$, $B = \begin{bmatrix} 1 & 2 \\ 0 & 1 \end{bmatrix}$ as two members of M, we find that

$$AB = \begin{bmatrix} 2 & 4 \\ 3 & 5 \end{bmatrix}\begin{bmatrix} 1 & 2 \\ 0 & 1 \end{bmatrix} = \begin{bmatrix} 2 & 8 \\ 3 & 11 \end{bmatrix} \text{ and } BA = \begin{bmatrix} 1 & 2 \\ 0 & 1 \end{bmatrix}\begin{bmatrix} 2 & 4 \\ 3 & 5 \end{bmatrix} = \begin{bmatrix} 8 & 14 \\ 3 & 5 \end{bmatrix}.$$

Thus, $AB \neq BA$ and so the ring is a non-commutative ring. But this ring is a ring with unity.

The unit matrix $I = \begin{bmatrix} 1 & 0 \\ 0 & 1 \end{bmatrix} \in M$ and is identity for multiplication of matrices because we have

$$IA = A = AI \ \forall \ A \in M.$$

Therefore the unit matrix I is the unity element of this ring.

Hence the ring of 2×2 matrices with elements as real numbers is a non-commutative ring with unity.

2. *Example of a commutative ring without unity:* The ring of even (2I, +, ·) *i.e.*, the ring

$$(\{..., -6, -4, -2, 0, 2, 4, 6,...\}, +, \cdot)$$

is a commutative ring without unity.

Since the multiplication of integers is a commutative operation, therefore the ring of even integers is a commutative ring. But this ring is without unity because it does not possess multiplicative identity. In the set of even integers there exists no even integer e such that

$$ea = a = ae, \ \forall \ a \in 2I.$$

Hence (2I, +, .) is a commutative ring without unity.

Example 26:

Explain with examples the difference between a field, a skew field and an integral domain.

Solution:

First give the definitions of a field, a skew field and an integral domain. For these definitions refer § 6.

Difference Between an Integral Domain and a Skew Field: A skew field need not be an integral domain because in a skew field multiplication need not be cumulative while in an integral domain multiplication must be commutative. Also an integral domain need not be a skew field. For example, the ring of integers is an integral domain but it is not a skew field.

Difference Between a Field and a Skew Field: Every field is a skew-field but a skew-field need not be a field. A skew field will be a field only if the operation of multiplication in it is commutative. (4) is not a field because the operation of multiplication of matrices in this skew field is not commutative.

Difference Between a Field and an Integral Domain: Every field is an integral domain but every integral domain need not be a field. For example the field of rational numbers (Q, +, .) is an integral domain because it does no possess zero divisors. The product of two non-zero rational numbers is never zero. On the other had, the ring of integers (I, +, ·) is an integral domain but it is not a field. The only inversible elements of this ring are 1 and – 1 while in a field every non-zero element must be inversible.

However, every finite integral domain is always a field. For example, the ring ({0, 1, 2}, $+_3$, ×3) ισ an integral domain and it is also a field.

Isomorphism of Rings

A ring R is said to be isomorphic to another ring S' if there exists a one-one mapping f of R onto R' such that

$$f\,(a+b) = f(a) + f(b),\ f(ab) = f(a)\ f(b)\ \ \forall\ a,\ b \in S$$

Also such a mapping f is said to be an isomorphism of R onto S'.

A mapping f : S → S' is an isomorphism of S onto S' if f is one-one and if f(a + b) = f(a) + f(b), f(ab) = f(a) f(b) ∀ a, b ∈ S. For f to be an isomorphism of R onto R', it must be one-one onto and it must preserve both addition and multiplication compositions in R and R'. There may be more than one isomorphisms of R onto R' if R is isomorphic to S'.

If a ring R is isomorphic to another ring R', we shall write in symbols S ≅ S'. Also R' is said to be an isomorphic image of S.

Example 27:

If in a ring with unity any element a has the multiplicative inverse, then a cannot be a divisor of zero.

Solution:

Let R be a ring with unity 1.

Let a ∈ R be such that a has multiplicative inverse. To prove that a cannot be a divisor of zero.

Let a ∈ R be such that ab = 0 ⇒ ba = 0. Then a cannot be a divisor of zero if we prove that ab = 0 ⇒ ba = 0 is possible only if b = 0

We have $ab = 0 \Rightarrow a^{-1}\,(ab) = a^{-1}0$ [$\because\ a^{-1}$ exists]

$$\Rightarrow (a^{-1}a)\,b = 0 \Rightarrow 1b = 0 \Rightarrow b = 0.$$

Again $ba = 0 \Rightarrow (ba)\,a^{-1} = 0a^{-1} \Rightarrow b(aa^{-1}) = 0 \Rightarrow b1 = 0 \Rightarrow b = 0.$

Hence a cannot be a zero divisor.

Example 28:

Prove that a ring R is commutative if and only if $(a + b)^2 = a^2 + 2ab + b^2 \ \forall \ a, b \in R$.

Solution:

Let R be a commutative ring *i.e.*, $ab = ba \ \forall \ a, b \in R$. Then to prove that $(a + b)^2 = a^2 + 2ab + b^2 \ \forall \ a, b \in R$. Let $a, b \in R$.

We have $(a + b)^2 = (a + b)(a + b)$ $\quad [\because \ a^2 = aa]$

$= aa + ab + ba + bb$, by dist, laws

$= a^2 + ab + ab + b^2$ $\quad [\because \ ba = ab]$

$= a^2 + 2ab + b^2$.

Hence if R is commutative ring, then

$(a + b)^2 = a^2 + 2ab + b^2 \ \forall \ a, b \in R$.

Conversely suppose that $(a + b)^2 = a^2 + 2ab + b^2 \ \forall \ a, b \in R$.

Then to prove that R is a commutative ring.

Let $a, b \in R$. Then

$(a + b)^2 = a^2 + 2ab + b^2$

$\Rightarrow$ $(a + b)(a + b) = a^2 + 2ab + b^2$

$\Rightarrow$ $a^2 + ab + ba + b^2 = a^2 + 2ab + b^2$, by dist. laws

$\Rightarrow$ $ab + ba = 2ab$, by left and right cancellation laws for addition in R

$\Rightarrow$ $ab + ba = ab = ab$ $\quad [\because \ ab + ab = 2ab]$

$\Rightarrow$ $ba = ab$, by left cancellation law for addition in R.

Thus $ab = ba \ \forall \ a, b \in R$ and so R is a commutative ring.

Example 29:

If R is a ring such that $a^2 = a \ \forall \ a \in R$ *prove that* (1) $a + a = 0 \ \forall \ a \in R$ *i.e., each element of R is its own additive inverse.*

(2) $a + b = 0 \Rightarrow a = b$. (3) R is a commutative ring.

Solution:

1. $a \in R \Rightarrow a + a \in R$.

Now $(a + a)^2 = (a + a)$ $\quad$ [given]

$\Rightarrow (a + a)(a + a) = a + a$

$\Rightarrow (a + a)a + (a + a)a = a + a$ $\quad$ [left dist. law]

$\Rightarrow (a^2 + a) + (a^2 + a^2) = a + a$ [right dist. law]

$\Rightarrow (a + a) + (a + a) = a + a$ [$\because$ $a^2 = a$]

$\Rightarrow (a + a) + (a + a) = (a + a) + 0.$ [$\because$ $a + 0 = a$]

$\Rightarrow (a + a) = 0$ [by left cancellation law for addition in R]

2. We have just proved that $a + a = 0$.

$\therefore$ $a + b\ 0 \Rightarrow a + b = a + a \Rightarrow b = a$, by left cancellation law for addition in R.

3. We have

$(a + b)^2 = (a + b)$

$\Rightarrow (a + b)(a + b) = (a + b)$

$\Rightarrow (a + b)\,a + (a + b)\,b = a + b$ [left dist. law]

$\Rightarrow (a^2 + ba) + (ab + b^2) = a + b$ [right dist. law]

$\Rightarrow (a + ba) + (ab + b) = a + b$ [$\because$ $a^2 = a, b^2 = b$]

$\Rightarrow (a + b) + (ba + ab) = (a + b) + 0$ [by commutativity and associativity of addition]

$\Rightarrow ba + ab = 0$ [by left cancellation law for addition in R]

$\Rightarrow ab = ba.$ [by part (2) of this question]

$\therefore$ R is commutative ring.

Note: An element a of a ring R is said to be idempotent if $a^2 = a$. *A ring R is called a* **Boolean Ring** *if all of its elements are idempotent i.e, if* $a^2 = a\ \forall\, a \in R$.

Example 30:

If R is a ring with unity element 1, then

$$(-1)\,a = -\,a = a\,(-1)\ \forall\, a \in R \text{ and } (-1)\,(-1) = 1.$$

Solution:

Let R be a ring with unity element 1. Let a be any element of R.

We have $0a = 0$

$\Rightarrow (-1 + 1)\,a = 0$ [$\because$ $-1 + 1 = 0$]

$\Rightarrow (-1)\,a + 1a = 0$ [by a dist. law]

$\Rightarrow (-1)\,a + a = 0$ [$\because$ $1a = a$]

$\Rightarrow (-1)\,a = -\,a.$

Again $a0 = 0$

$\Rightarrow a(-1+1)=0$

$\Rightarrow a(-1)+a1=0$

$\Rightarrow a(-1)+a=0$

$\Rightarrow a(-1)=-a.$

Hence $(-1)a=-a=a(-1)\ \forall\ a\in R$.

Now $(-1)(-1)=-(-1)$ $\quad$ [$\because$ $(-1)=-a$, as proved above]

$=1.$ $\quad$ [$\because$ in a ring, $-(-a)=a$]

Example 31:

Show that the set of numbers of the form $a+b\sqrt{2}$, with a and b as rational numbers is a field.

Solution:

Let $R=\{a+b\sqrt{2} : a, b, \in Q\}$.

Let $a_1+b_1\sqrt{2}\in R$ and $a_2+b_2\sqrt{2}\in R$. Then

$$a_1, b_1, a_2, b_2 \in Q.$$

We have $(a_1+b_1\sqrt{2})+(a_2+b_2\sqrt{2})=(a_1+a_2)+(b_1+b_2)\sqrt{2}$

$\in R$ since $a_1+a_2, b_1+b_2\in Q$.

Also $(a_1+b_1\sqrt{2})+(a_2+b_2\sqrt{2})=(a_1a_2+2b_1b_2)+(a_1b_1+a_2b_1)\sqrt{2}$

$\in R$ since $a_1a_2+2b_1b_2, a_1b_2+a_2b_1\in Q$.

The R is closed with respect to addition and multiplication.

All the elements of R are real numbers and we know that addition and multiplication are both associative as well as commutative compositions in the set of real numbers.

Further we have $0+0\sqrt{2}\in R$ since $0\in R$

If $a+b\sqrt{2}\in R$, then

$$(0+0\sqrt{2})+(a+b\sqrt{2})=(0+a)+(0+b)\sqrt{2}=a+b\sqrt{2}.$$

$\therefore 0+0\sqrt{2}$ is the *additive identity*.

Again if $a+b\sqrt{2}\in R$, then $(-a)+(-b)\sqrt{2}\in R$ and we have

$$[(-a)+(-b)\sqrt{2}]+(a+b\sqrt{2})=0+0\sqrt{2}.$$

$\therefore$ Each element of R possesses additive inverse.

Further in the set of real numbers multiplication is distributive with respect to addition.

Again $1+0\sqrt{2}\in R$ and we have

$$(1 + 0\sqrt{2})(a + b\sqrt{2}) = a + b\sqrt{2} = (a + b\sqrt{2})(1 + 0\sqrt{2}).$$

$\therefore 1 + 0\sqrt{2}$ is the multiplicative identity.

Thus R is a commutative ring with unity. The zero element of the ring is $0 + 0\sqrt{2}$ and the unit element is $1 + 0\sqrt{2}$.

Now R will be a field, if each non-zero element of R possesses multiplicative inverse.

Let $a + b\sqrt{2}$ be any non-zero element of this ring *i.e.* at least one of a and b is not zero.

Then $$\frac{1}{a + b\sqrt{2}} = \frac{a - b\sqrt{2}}{\left(a + b\sqrt{2}\right)\left(a - b\sqrt{2}\right)} = \frac{a - b\sqrt{2}}{a^2 - 2b^2}$$

$$= \left(\frac{a}{a^2 - 2b^2}\right) + \left(-\frac{b}{a^2 - 2b^2}\right)\sqrt{2}.$$

Now if a and b are rational numbers, then we can have $a^2 = 2b^2$ only if $a = 0$, $b = 0$. Since here at least one of the rational numbers a and b is not 0, therefore we cannot have $a^2 = 2b^2$ *i.e.*, $a^2 - 2b^2 = 0$.

$\therefore \dfrac{a}{a^2 - 2b^2}$ and $-\dfrac{b}{a^2 - 2b^2}$ are both rational numbers and at least one of them is not zero.

$\therefore \left(\dfrac{a}{a^2 - 2b^2}\right) + \left(-\dfrac{b}{a^2 - 2b^2}\right)\sqrt{2}$ is a non-zero element of R and is the multiplicative inverse of $a + b\sqrt{2}$.

Hence the given system is a field.

Example 32:

Show that the set R of all real valued continuous functions defined in the closed interval [0, 1] is a commutative ring with unity with respect to the addition and multiplication of functions defined pointwise as follows:

$$(f + g)(x) = f(x) + g(x)$$

and $$(fg)(x) = f(x)\, g(x),$$

where f, g are any two members of R.

Solution:

If f is a real valued function in the closed interval [0, 1], then we mean that f (x) is a real number $\forall\ x \in [0, 1]$. We know that the sum and product of two real numbers are also real numbers. Also the sum and product of two continuous functions is also a continuous function. Therefore R is closed with respect to the given compositions.

Now let f, g, h b any three elements of R. Then we make the following observations:

Associativity of Addition: For every $x \in [0, 1]$, we have

$[(f + g) + h] (x) = [(f + g) (x)] + h(x)$

$= [(f(x) + g(x)] + h(x) = f(x) + [g(x) + h(x)]$

[$\because$ f(x), g(x), h(x) are real numbers and addition of real numbers is associative]

$= f(x) + (g + h) (x) = [f + (g + h)] (x).$

$\therefore (f + g) + h = f + (g + h)$, by the equality of two mappings.

Commutativity of Addition: We have

$(f + g) (x) = f(x) + g(x) = g(x) + f(x)$

[$\because$ addition of real numbers is commutative]

$= (g + f) (x).$

$\therefore \quad f + g = g + f.$

Existence of Additive Identity: Let us define a function e by the rule $e(x) = 0 \ \forall \ x \in [0, 1]$. Then $e \in R$. Also if $f \in R$. we have $(e + f)(x) = e(x) + f(x) = 0 + f(x) = f(x)$.

Existence of Additive Inverse: Let $f \in R$. Let us define a function $-f$ by the formula $(-f)(x) = -[f(x)] \ \forall \ x \in [0, 1]$.

Then $-f \in R$ and we have

$[-f + f] (x) = (-f) (x) + f(x) = - f(x) + f(x) = 0 = e(x).$

$\therefore \quad -f + f = e.$

$\therefore$ The function $-f$ is the additive inverse of f.

Associativity of Multiplication: We have

$[(fg) h] (x) = [(fg) (x)] h(x)$

$= [f(x) g(x)] h(x) = f(x) [g(x) h(x)]$

$= f(x) [(gh) (x)] = [f(gh)] (x).$

$\therefore \quad (fg) h = f(gh).$

Distributive Laws: We have

$[f (g + h) (x) = f(x) [(g + h) (x)] = f(x) [g(x) + h(x)]$

$= f(x) g(x) + f(x) h(x) = (fg) (x) = (fh) (x)$

$= [fg + fh] (x).$

$\therefore \quad f(g + h) = fg + fh.$

Similarly we can prove that $(g + h) f = gf + hf$.

$\therefore$ R is a ring with respect to the given compositions.

Commutativity of Multiplication: We have

$(fg) (x) = f(x) g(x) = g(x) f(x) = (gf) (x).$

$\therefore \quad fg + gf.$

$\therefore$ R is a commutative ring.

Existence of Multiplicative Identity: Let the define a function i by the formula $i(x) = 1 \ \forall \ x \in [0, 1]$. Then $i \in R$. If $f \in R$, we have $(if) (x) = i(x) f(x) = 1 \ f(x) = f(x)$.

$\therefore if + f = fi$ [by commutativity of multiplication]

$\therefore$ The function i is the multiplicative identity

Thus, the ring R is with unity element.

Example 33:

If addition and multiplication modulo 10 is defined on the set of integers $R = \{0, 2, 4, 6, 8\}$, prove that the resulting system is a ring with unity. Is it an integral domain ?

Solution:

We have $R = \{0, 2, 4, 6, 8\}$. To prove that $(R, +_{10}, \times_{10})$ is a ring with unity.

First we shall show that $(R, +_{10})$ is an abelian group. The composition table for R for the operation $+_{10}$ is as given below.

$+_{10}$	0	2	4	6	8
0	0	2	4	6	8
2	2	4	6	8	0
4	4	6	8	0	2
6	6	8	0	2	4
8	8	0	2	4	6

We see that, all the entries in the composition table are elements of the set R. Therefore R is closed with respect to $+_{10}$

The composition $+_{10}$ on the set R is commutative. For if a, b, $\in$ R, then

$a +_{10} b$ = least non-negative remainder when a + b is divided by 10

= least non-negative remainder when b + a is divided by 10

$= b +_{10} a$.

The composition $+_{10}$ on the set R is associative. For if a, b, c $\in$ R, then

$a +_{10} (b +_{10} c) = (a +_{10} b) +_{10} c$ because $a + (b + c) = (a + b) + c$.

Existence of Additive Identity: We have $0 \in R$

and $$0 +_{10} a = a = a +_{10} 0 \ \forall \ a \in R.$$

$\therefore$ 0 is the identity for the operation $+_{10}$ and will be the zero element of the ring $(R, +_{10}, \times_{10})$.

Existence of Additive Inverse: From the composition table we see that the inverses of 0, 2, 4, 6, 8 for the operation $+_{10}$ are 0, 8, 6, 4, 2 respectively. Thus each element of R possesses inverse for $+_{10}$.

$\therefore (R, +_{10})$ is an abelian group.

Now we prepare the composition table for R for the operation $\times_{10}$.

$+_{10}$	0	2	4	6	8
0	0	0	0	0	0
2	0	4	8	2	6
4	0	8	6	4	2
6	0	2	4	6	8
8	0	6	2	8	4

We see that all, the entries in the composition table are elements of the set R. Therefore R is closed with respect to $\times_{10}$.

The operation $\times_{10}$ on the set R is associative. For if a, b, c $\in$ R, then

$a \times_{10}(b \times_{10} c) = (a \times_{10} b) \times_{10} c$ because $a (bc = (ab) c$.

Also the operation $\times_{10}$ distributes over the operation $\times_{10}$. For if a, b, c $\in$ R, then

$$a \times_{10}(b \times_{10} c) = (a \times_{10} b) \times_{10} (a \times_{10} c)$$

and $$(b \times_{10} c) \times_{10} a = (b \times_{10} a) +_{10} (c \times_{10} a).$$

$\therefore$ The algebraic structure $(R, +_{10}, \times_{10})$ is a ring. The zero element of this ring to 0 which is identity for the operation $+_{10}$.

This ring is a ring with unity because it possesses multiplicative identity. From the composition table for $\times_{10}$ we see that $6 \in R$ is identity for the operation of multiplication modulo 10.

We have

$$6 \times_{10} a = a = a \times_{10} 6, \forall a \in R.$$

Thus, 6 is identity for the operation $\times_{10}$ and is therefore the unity element of the ring $(R, +_{10} \times_{10})$.

The operation $\times_{10}$ on the set R is commutative because the corresponding rows and columns in the composition table of R for the operation $\times_{10}$ are identical. Thus

$$a \times_{10} b = b \times_{10} a, \forall a, b \in R.$$

Also from the composition table of R for the operation $\times_{10}$ we observe that the product of no two non-zero elements of R is the zero element or R. Thus if a, b, $\in$ R, then

$$a \times_{10} b = 0 \Rightarrow a = 0 \text{ or } b = 0.$$

$\therefore$ The ring $(R, +_{10} \times_{10})$ does not possess zero divisors.

Since $(R, +_{10} \times_{10})$ is a commutative ring with unity and without zero divisors, therefore it is an integral domain.

Example 34:

Let C be the set of the ordered pairs (a, b) of real numbers. Define addition and multiplication in C by the equations

$$(a, b) + (c, d) = (a + c, b + d)$$

$$(a, b)\ (c, d) = (ac - bd, bc + ad).$$

Prove that C is a field.

Solution:

We see that C is closed with respect to the two compositions since $a + c$, $b + d$, $ac - bd$, $bc + ad$ are all real numbers. Now let (a, b), (c, d), (e, f) be any elements of C. Then we make the following observations:

Associativity of Addition: We have

$$[(a, b) + (c, d)] + (e, f) = (a + c, b + d) + (e, f)$$

$$= ([a + c] + e, [b + d] + f) = (a + [c + e], b + [d + f])$$

$$= (a, b) + (c + e, d + f) = (a, b) + [(c, d) + (e, f)].$$

Commutativity of Addition: We have

$(a, b) + (c, d) = (a + c, b + d) = (c + a, d + b) = (c, d) + (a, b)$.

Existence of Additive Identity: We have $(0, 0) \in C$.

Also $(0, 0) + (a, b) = (0 + a, 0 + b) = (a, b)$.

$\therefore (0, 0)$ is the additive identity.

Existence of Additive Inverse: If $(a, b) \in C$, then

$(-a, -b) \in C$. We have

$$(-a, -b) + (a, b) = (-a + a, -b + b) = (0, 0).$$

$\therefore (-a, -b)$ is the additive inverse of (a, b).

Associativity of Multiplication: We have

$[(a, b)(c, d)](e, f) = (ac - bd, bc + ad)(e, f)$

$= ([ac - bd] e - [bc + ad] f, [bc = ad] e + [ac - bd] f)$

$= (a [ce - df] - b [de + ef], b [ce - df] + a [de + ef])$

$= (a, b)(ce - df, de + cf) = (a, b)[(c, d)(e, f)]$.

Distributive Laws: We have

$(a, b)[(c, d) + (e, f)] = (a, b)(c + e, d + f)$

$= (a [c + e] - b [d + f], b [c + e] + a [d + f])$

$= ([ac - bd] + [ae - bf], [bc + ad] + [be + af])$

$= (ac - bd, bc + ad) + (ae - bf, be + af) = (a, b)(c, d) + (a, b)(e, f)$.

Similarly we can show that the other distributive law also holds good.

Therefore C is a ring with respect to the two compositions. The ordered pair $(0, 0)$ is the zero element of the ring.

Commutativity of Multiplication: We have

$(a, b)(c, d) = (ac - bd, bc + ad) = (ca - db, da + cb) = (c, d)(a, b)$.

Existence of Multiplicative Identity: We have $(1, 0) \in C$. If $(a, b) \in C$, then $(a, b)(1, 0) = (a1 - b0, b1 + a0) = (a, b) = (1, 0)(a, b)$. Therefore $(1,0)$ is the unity element of the ring.

Existence of multiplicative inverse of each non-zero element of c. Let (a, b) be any non-zero element of C. Then a and b are not both simultaneously zero. If (c, d) is the multiplicative inverse of (a, b), then we should have

$$(a, b)(c, d) = (1, 0)$$

$$\Rightarrow \quad (ac - bd, bc + ad) = (1, 0).$$

By the definition of the equality of two ordered pairs, we have

$$ac - bd = 1 \text{ and } bc + ad = 0.$$

Solving these equations for c, d, we get

$$c = \frac{a}{a^2 + b^2}, d = \left(-\frac{b}{a^2 + b^2}\right).$$

Now $a \neq 0$ or $b \neq 0 \Rightarrow a^2 + b^2 \neq 0$. Therefore either c or d or both are non-zero real numbers. Thus $\left(\frac{a}{a^2 + b^2}, -\frac{b}{a^2 + b^2}\right)$ is the multiplicative inverse of (a, b). Hence c is a field.

Note: In this question C is nothing but the set of complex numbers defined as ordered pairs of real numbers. Thus, we have proved that the set of complex numbers is a field with respect to addition and multiplication of complex numbers.

Example 35:

Give an example of a skew field which is not a field.

Solution:

Let M be the set of all 2 × 2 matrices of the form

$$\begin{bmatrix} a + ib & c + id \\ -c + id & a - ib \end{bmatrix},$$

where a, b, c, d are arbitrary real numbers.

First we shall prove that M is a ring with respect to addition and multiplication of matrices.

Let $$A = \begin{bmatrix} a + ib & c + id \\ -c + id & a - ib \end{bmatrix} \text{ and } B = \begin{bmatrix} p + iq & r + is \\ -r + is & p - iq \end{bmatrix}$$

be any two elements of M. We have

$$A + B = \begin{bmatrix} (a + p) + i(b + q) & (c + r) + i(d + s) \\ -(c + r) + i(d + s) & (a + p) - i(b + q) \end{bmatrix},$$

which is obviously a matrix of the given form. So $A + B \in M$.

Also AB

$$= \begin{bmatrix} (a+ib)(p+iq) + (c+id)(-r+is) & (a+ib)(r+is) + (c+id)(p-iq) \\ (-c+id)(p+iq) + (a-ib)(-r+is) & (-c+id)(r+is) + (a+ib)(p-iq) \end{bmatrix}$$

$$= \begin{bmatrix} (ap - bq - cr - ds) & (ar - bs + cp + dq) \\ \quad + i\,(aq + bp + cs - dr) & \quad + i\,(as + br - cq + dp) \\ -(ap + dq + ar - bs) & (-cr - ds + ap - bq) \\ \quad + i\,(dp - cq + as + br) & \quad - i\,(cs - dr + aq + bp) \end{bmatrix}$$

which is obviously an element of M. Thus M is closed with respect to addition and multiplication of matrices. Further matrix addition is commutative as well as associative. The zero matrix $\begin{bmatrix} 0 + i0 & 0 + i0 \\ -0 + io & 0 - i0 \end{bmatrix}$ is the additive identity and so it is the zero element of M. If $A = \begin{bmatrix} a + ib & c + id \\ -c + id & a - ib \end{bmatrix} \in M$, then obviously $-A = \begin{bmatrix} -a - ib & -c + id \\ c - id & -a + ib \end{bmatrix} \in M$. Thus each element of M possesses additive inverse. Further matrix multiplication is associative and it is distributive with respect to addition. Hence M is a ring with respect to addition and multiplication of matrices.

Existence of multiplicative identity: The matrix

$\begin{bmatrix} 1 + i0 & 0 + 0i \\ -0 + i0 & 1 - i0 \end{bmatrix} = \begin{bmatrix} 1 & 0 \\ 0 & 1 \end{bmatrix}$ is obviously an element of M. It is the unit matrix and so it is the multiplicative identity. Thus M is a ring with unity.

Existence of multiplicative inverse of each non-zero element of M.

Let $A = \begin{bmatrix} a + ib & c + id \\ -c - id & a - ib \end{bmatrix}$ be any non-zero element of M *i.e.*, a, b, c, d are not all equal to zero. We have $|A|$ *i.e.*, det. $A = a^2 + b^2 + c^2 + d^2 \neq 0$. Therefore the matrix A is non-singular and is therefore inversible We must show that $A^{-1} \in M$. Let $|A| = m$. Then m is a real number and $m > 0$. We have

$$A^{-1}\ 0\ \frac{1}{|A|}\ \text{Adj. } A = \frac{1}{m}\begin{bmatrix} a - ib & -c - id \\ c - id & a + ib \end{bmatrix}$$

which is obviously an element of M. Therefore each non-zero element of M possesses multiplicative inverse.

$\therefore$ M is a skew field (or a division ring).

Now we shall show that M is not a field *i.e.*, multiplication in M is not commutative. We have $A = \begin{bmatrix} 3 + 4i & 5 + 6i \\ -5 + 6i & 3 - 4i \end{bmatrix} \in M$, and $B = \begin{bmatrix} 1 + i0 & 1 + i0 \\ -1 + i0 & 1 - i0 \end{bmatrix} = \begin{bmatrix} 1 & 1 \\ -1 & 1 \end{bmatrix} \in M.$

Also $AB = \begin{bmatrix} -2 - 2i & 8 + 10i \\ -8 + 10i & -2 + 2i \end{bmatrix}$, and $BA = \begin{bmatrix} -2 - 10i & 8 + 2i \\ -8 + 2i & -2 + 10i \end{bmatrix}$.

We see that, $AB \neq BA$. Therefore multiplication in M is not commutative and so M is not a field.

Hence M is a skew-field which is not a field.

Example 36:

Prove that in general the set of all numbers of the form

$$a + (\sqrt{p})\, b$$

where a, b are rational numbers and p is a prime is a field with respect to ordinary addition and multiplication.

Solution:

We can easily verify that the given system is a commutative ring with unit element $1 + \sqrt{p}0$. Also $0 + \sqrt{p}0$ is the zero element of this ring. Further let $a + \sqrt{p}b$ be any non-zero element of the given set.

Then $\dfrac{1}{a + \sqrt{p}b} = \dfrac{a - \sqrt{p}b}{a^2 - pb^2} = \dfrac{a}{a^2 - pb^2} + \left(-\dfrac{b}{a^2 - pb^2}\right)\sqrt{p}.$

Now if p is prime and if at least one of the rational numbers a and b is not zero, then we cannot have $a^2 = pb^2$.

$\therefore \dfrac{a}{a^2 - pb^2}$ and $-\dfrac{b}{a^2 - pb^2}$ are both rational numbers and at least one them is not zero.

$\therefore \left(\dfrac{a}{a^2 - pb^2}\right) + \left(\dfrac{b}{a^2 - pb^2}\right)\sqrt{p}$ is a non-zero element of the given set and it is the multiplicative inverse of the non-zero element $a + b\sqrt{p}$. Hence the given system is a field.

Example 37:

Prove that the totality R of all ordered pairs (a, b) of real numbers is a commutative ring with zero divisors under the addition and multiplication of ordered pairs defined as

$$(a, b) + (c, d) = (a + c, b + d)$$
$$(a, b) + (c, d) = (ac, bd)$$
$$\forall (a, b) + (c, d) \in R.$$

Solution:

We see that R is closed with respect to the two compositions since a + c, b + d, ac, bd are all real numbers. Now let (a, b), (c, d), (e, f) be any elements of R. Then we observe:

Associativity of Addition: We have

$[(a, b) + (c, d)] + (e, f) = (a + c, b + d) + (e, f)$

$= ([a + c] + e, [b + d] + f)$

$= (a + [c + e], b + [d + f])$

[$\because$ addition of real numbers is associative]

$= (a, b) + (c + e, d + f)$

$= (a, b) + [(c, d) + (e, f)].$

$\therefore$ addition in R is associative.

Commutativity of Addition: We have

$(a, b) + (c, d) = (a + c, b + d) = (c + a, d + b) = (c, d) + (a, b).$

Existence of Additive Identity: We have $(0, 0) \in R$, Also $(0, 0) + (a, b) = (0 + a, 0 + b) = (a, b)$.

Existence of Additive Inverse: If $(a, b) \in R$, then $(-a, -b) \in R$ and we have $(-a, -b) + (a, b) = (-a + a, -b + b) = (0, 0)$.

$\therefore$ $(-a, -b)$ is the additive inverse of (a, b)

Associativity of Multiplication: We have

$[(a, b) (c, d)] (e, f) = (ac, bd) (e, f) = ([ac] e, [bd] f)$

$= (a [ce], b [df]).$

[$\because$ multiplication of real numbers is associative]

$= (a, b) (ce, df) = (a, b) [(c, d) (e, f)].$

Distributive Laws: We have

$(a, b) [(c, d) + (e, f)] = (a, b) (c + e, d + f)$

$= (a [c + e], b [d + f])$

$= (ac + ae, bd + bf) = (ac, bd) + (ae, bf)$

$= (a, b) (c, d) + (a, b) (e, f).$

Similarly we can show that the other distributive law also holds good.

∴ R is a ring with respect to the given compositions.

Commutativity of Multiplication: We have

(a, b) (c, d) = (ac, bd) = (ca, db) = (c, d) (a, b).

∴ R is s commutative ring.

Existence of Multiplicative Identity: We have (1, 1) ∈ R. If (a, b) ∈ R, then (1, 1) (a, b) = (1a, 1b) = (a, b) = (a, b) (1, 1).

∴ (1, 1) is the multiplicative identity and is therefore the unit element of the ring. So r is a ring with unity also.

The zero element of this ring is the ordered pair (0, 0).

Now in order to show that R is a ring with zero divisors we should show that there exist two non-zero elements of R whose product is equal to the zero element of R. Obviously neither (3, 0) nor (0, 5) is equal to the zero element of R. But (3, 0) (0, 5) = (3 × 0, 0 × 5) = (0, 0) which is the zero element of R.

∴ R is a ring with zero divisors.

Example 38:

Give an example each of:

1. *a commutative ring without unity,*
2. *a non-commutative ring,*
3. *a ring without zero divisors,*
4. *division ring.*

Solution:

1. A commutative ring without unity.
2. A non-commutative ring
3. A ring without zero divisors. The ring of integers (I, +, .) is a ring without zero divisors. We know that the product of two non-zero integers is never zero. Thus if a, b, ∈ I, then

   $$ab = 0 \Rightarrow a = 0 \Rightarrow b = 0.$$

   Hence the ring of integers is a ring without zero divisors.
4. Division ring. Let M be the set of all 2 × 2 matrices of the form

   $$\begin{bmatrix} a + ib & c + id \\ -c + id & a - ib \end{bmatrix},$$

where a, b, c, d are arbitrary real numbers.

Then M is a ring for addition and multiplication of matrices as the two ring operations. This ring possesses unity element.

The unit matrix $I = \begin{bmatrix} 1 & 0 \\ 0 & 1 \end{bmatrix} \in M$ and we have

$$IA = A = AI, \forall A \in M.$$

Also in this ring every non-zero element possesses multiplicative inverse. But the operation of multiplication of matrices on the set M is not commutative. Hence the above ring M is a division ring or a skew field.

Example 39:

Prove that the set of integers is on integral domain with respect to addition and multiplication.

Solution:

Let I = {..., – 3, – 2, – 1, 0, 1, 2, 3,...} be the set of integers. To prove that the algebraic structure (I, +, .) is an integers is also an integer. Therefore I is closed for addition of integers.

Also addition of integers is commutative as well as associative. The integer 0 is identity for addition of integers. Also if a is any integer in I, then the integer –a is its additive inverse. Thus (I, +) is an abelian group.

Now the product of two integers is also an integer. Therefore I is closed for multiplication of integers.

Also multiplication of integers is associative as well as commutative and it distributes over addition of integers. The integer 1 is identity for multiplication. Further the product of two integers can be zero only if at least one of them is zero.

Hence the algebraic structure (I, +, .) is a commutative ring with unity and without zero divisors and so it is an integral domain.

Example 40:

Prove that the set of rational number (real numbers or complex numbers) is a field with respect to addition and multiplication.

Solution:

We shall give proof in the case of the set of rational numbers and a similar proof can be given in these of the set of real numbers or the set of complex numbers.

Let Q denote the set of rational numbers. Remember that a rational number is of the form a/b, where a and b are integers and b ≠ 0.

To prove that the algebraic structure (Q, +, .) is a field

(Q, +) is an abelian group: We know that the sum of two rational numbers is also a rational number. Therefore Q is closed for addition of rational numbers.

Also addition of rational numbers is commutative as well as associative. The rational number 0 is identity for addition of rational numbers. Also if a/b is its additive inverse. Thus (Q, +) is an abelian group.

Now the product of two rational numbers is also a rational number. Therefore Q is closed for multiplication of rational numbers.

Also multiplication of rational numbers is commutative as well as associative and it distributes over addition of rational numbers. The rational number 1 is identity for multiplication. Thus (Q, +, .) is a commutative ring with unity 1.

If a/b is any non-zero rational number, then the integer a ≠ 0 and so b/a is also a rational number.

We have (a/b) (b/a) = 1, so that b/a is the multiplicative inverse of a/b. Thus each non-zero rational number possesses multiplicative inverse. Hence (Q, +, .) is a field.

Remark: *To give the proof in the case of the set of complex numbers denote the set of the complex numbers by C.*

Remember that a complex number is of the form a + ib where a and b are any real numbers.

If a + ib and c + id are any two complex numbers, then

(a + ib) + (c + id) = (a + c) + i (b + c) which is also a complex number

and (a + ib) (c + id) = (a + c) + i (b + d) which is also a complex number.

The complex number 0 *i.e.*, 0 + i0 is identity for addition of complex numbers. Also if a + ib is any complex number, then the complex number –a – ib *i.e.*, (–a) + i (–b) is its additive inverse.

The complex number 1 *i.e.*, 1 + i0 is identity for multiplication.

Finally every non-zero complex number possesses multiplicative inverse as shown below.

Let a + ib be any non-zero complex number *i.e.*, a + ib ≠ 0 + i0. Then a and b are real numbers and at least one of them is not zero.

Suppose the complex number x + iy is the multiplicative inverse of a + ib. Then

$$(x + iy)(a + ib) = 1 + i0$$

$$\Rightarrow (xa - yb) + i(xb + ya) = 1 + i0$$

$$\Rightarrow xa - yb = 1,\ xb + ya = 0.$$

Solving the equations xa − yb = 1, xb + ya = 0, we get

$$x = \frac{a}{a^2 + b^2},\ y = \frac{-b}{a^2 + b^2}.$$

Since at least one of a and b is not zero, therefore the real number $a^2 + b^2 \neq 0$ and so both x and y are some real numbers.

Therefore the complex number $\frac{a}{a^2 + b^2} + i\left(\frac{-b}{a^2 + b^2}\right)$ is the multiplicative inverse of the non-zero complex number a + ib.

1.10 RELATION OF ISOMORPHISM IN THE SET OF ALL RINGS

Example 1:

Let R be the ring of integers under ordinary addition and multiplication. Let S' be the set of all even integers. Let us define multiplication in S' to be denoted by '' by the relation*

$$a * b = \frac{ab}{2},$$

where ab is the ordinary multiplication of two integers a and b.

1. *Prove that (S, +, ∗) is a commutative ring where + stands for ordinary addition of integers.*
2. *Prove that S is isomorphic to S'*
3. *What acts as the unit element of S' ?*

Solution:

Obviously S' is an abelian group with respect to addition.

If a and b are both even integers then $\frac{ab}{2}$ is also an even integer.

Therefore $a * b = \left(\frac{ab}{2}\right) \in S'\ \forall\ a, b, \in S'$.

Thus S' is closed with respect to ∗.

Also if a, b, c are any elements of R', then

$$a * (b * c) = a * \left(\frac{ab}{2}\right) = \frac{a\,(bc/2)}{2} = \frac{(ab/2)\,c}{2} = \left(\frac{ab}{2}\right) * c = (b * c) * c.$$

$\therefore$ $*$ is associative.

Further $(a * b) = \dfrac{ab}{2} = \dfrac{ba}{2} = (b * a)$.

$\therefore$ $*$ is commutative.

Again $a * (b + c) = \dfrac{a(b+c)}{2} = \dfrac{ab}{2} + \dfrac{ac}{2} = (a * b) + (a * c)$.

Similarly $(b + a) * a = (b * a) + (c * a)$.

$\therefore$ $*$ is distributive with respect to +.

$\therefore$ (S, +, $*$) is a commutative ring.

2. We shall now show that S is isomorphic to S'. Consider the mapping f defined by

   $f : S \to S'$ such that $f(x) = 2x \;\forall\; x \in S$.

The mapping f is obviously one-one onto.

Also $\forall\; x_1, x_2 \in S$ we have

$$f(x_1\; x_2) = 2\,(x_1 + x_2) = 2x_1 + 2x_2 = f(x_1) + f(x_2)$$

and $f(x_1 x_2) = 2\,(x_1 x_2) = \dfrac{(2x_1)\,(2x_2)}{2} = (2x_1) * (2x_2) = f(x_1) * f(x_2)$.

$\therefore$ The mapping f is an isomorphism of S onto S'.

3. Here 1 is the unit element of S. We have f(1) = 2. Since f is an isomorphism of S onto S', therefore 2 must be the unity element of R'. We have for all $a \in S'$, $2 * a = \dfrac{2a}{2} = a = a * 2$.

$\therefore$ 2 is the unit element of S'.

1.11 PROPERTIES OF ISOMORPHISM OF RINGS

Theorem:

If f is an isomorphism of a ring S onto a ring S', then

1. *The image of the zero of S is the zero of S'.*
2. *The image of the negative of an element of S is the negative of the image of that element i.e,* $f(-a) = -f(a) \;\forall\; a \in S$.
3. *If S is a commutative ring, then S' is also a commutative ring.*

4. *If S is without zero divisors, then S' is also without zero divisors.*
5. *If S is with unit element, then S' is also with unity element.*
6. *If S is a field, then S' is also a field.*
7. *If S is a skew field, then S' is also a skew field.*

Proof:

1. Let a ∈ S. Then f (a) ∈ S'. Let 0 denote the zero element of S'. To prove that f (0) = 0.

   We have f (a) 0' = f (a) = f (a + 0) = f (a) + f (0). By cancellation law for addition in R', we get from f (a) + 0' = f (a) + f (0), the result that 0' = f (0).

2. We have f (a) + f (–a) = f [a + (–a)] = f (0) = 0'.

∴ f (a) is the additive inverse of f(a) in S'. Thus

f(–a) = – f(a).

3. Let f (a) and f (b) be any two elements of S'. Then

   a, b ∈ S.

   we have f (a) f (b) = f (ab) = (ba)

   [∵ R is commutative ⇒ ab = ba]

   = f (b) f (a).

∴ S' is also commutative.

4. We have f (0) = 0'. Also f is one-one. Therefore 0 is the only element of S whose f-image is 0'.

Let f (a), f (b) be two non-zero elements of S'. Then f (a) ≠ 0', f (b) ≠ 0' ⇒ a ≠ 0, b ≠ 0. Since S is without zero divisors, therefore

a ≠ 0, b ≠ 0 ⇒ ab ≠ 0, ⇒ f (ab) ≠ f (0)

⇒ f (a) f (b) ≠ 0' ⇒ S' is without zero divisors.

5. Let 1 be the unit element of S. Then f (1) ∈ S'. If f (a) is any element of S', we have

   f (1) f (a) = f (1a) = f (a) and f (a) f(1) = f (a1) = f (a).

   ∴ f (1) is the unit element of S'.

6. If r is a field, then S is commutative, with unity and each non-zero element of S will possess multiplicative inverse. Now as proved in (3) and (5), S' will be commutative and will also have the unit element *i.e.*, f(1).

Let f (a) be any non-zero element of S'. Then

$f(a) \neq 0' \Rightarrow a \neq 0 \Rightarrow a^{-1}$ exists.

Now $f(a^{-1}) \in S'$ and we have

$f(a^{-1}) f(a) = f(a^{-1} a) = f(1)$ and $f(a) f(a^{-1}) = f(aa^{-1}) = f(1)$.

$\therefore$ $f(a^{-1})$ is the multiplicative inverse of f (a).

Hence S' is a skew-field.

1.12 TRANSFERENCE OF RING STRUCTURE

Subrings

Definition: *Let R be a ring. A non-empty subset S of the set R is said to be a subring of R if S is closed with respect to the operations of addition and multiplication in R and S itself is a ring for these operations.*

If S is a subring of a ring R, it is obvious that S is a subgroup of the additive group of R.

If R is any ring, then {0} and k itself are always subrings of R. These are know·as *Improper subrings* of R. Other subrings, if any, of R are called *proper subrings* of R.

Theorem 1:

If f is a one-one mapping of a ring R onto a set R' with two compositions denoted additively and multiplicatively such that $f(a + b) = f(a) + f(b)$, $f(ab) = f(a) f(b)$ $\forall$ a, b $\in$ R, then the set R' is a ring for the two compositions.

Conditions for a Subring

Theorem 2:

The necessary and sufficient conditions for a non-empty subset S of a ring R to be a subring of R are

1. $a \in S, b \in S \Rightarrow a - b \in S$
2. $a \in S, b \in S \Rightarrow ab \in S$.

Proof:

The conditions are necessary. Suppose (S, +, ·) is a subring of (R, +,·).

Since S is a group with respect to addition, therefore b = $\in$ S

$\Rightarrow -b \in S$.

Now S is closed with respect to addition.

$\therefore a \in S, b \in S \Rightarrow a \in S, -b \in S$

$$\Rightarrow a + (-b) \in S \Rightarrow a - b \in S.$$

Also S is closed with respect to multiplication.

$\therefore a \in S, b \in S \Rightarrow ab \in S.$

Hence the conditions are necessary.

The conditions are sufficient: Suppose S is a non-empty subset of R and the conditions (1) and (2) are satisfied. From (1), we have

$$a \in S, a \in S \Rightarrow a - a \in S$$

$$\Rightarrow 0 \in S \text{ i.e., the zero element} \in S.$$

Now since $0 \in S$, therefore from (1), we have

$0 \in S, a \in S \Rightarrow 0 - a \in S \Rightarrow -a \in S$ *i.e.*, each element of S Possesses additive inverse.

Now if a, b are any elements of S, then $-b \in S$.

From (1), we have

$$a \in S, -b \in S \Rightarrow a (-b) \in S \Rightarrow a + b \in S.$$

$\therefore$ S is closed with respect to addition.

Now, S is a subset of R. Therefore associativity and commutativity of addition must hold in S since they hold in R.

$\therefore$ (S, +) is an abelian group.

From (2) S is closed with respect to multiplication.

Associativity of multiplication and distributivity of multiplication over addition must hold in S since they hold in R.

Hence S is a subring of R.

Cor. : *The necessary and sufficient conditions for a non-empty subset S of a ring R to be a subring of R are*

(1) $S + (-S) = S$ (2) $SS \subseteq S$.

Proof:

The conditions are necessary: Suppose S is a subring of R. Then S is a subgroup of the additive group of R.

Let $a + (-b)$ be any element of $S + (-S)$.

We have

$a + (-b) \in S + (-S) \Rightarrow a \in S, -b \in -S$

$$\Rightarrow a \in S, b \in S$$

$\Rightarrow a - b \in S$ [$\because$ S is a subgroup]

$\therefore \quad S + (-S) \subseteq S.$

Also let a be any element of S. We can write $a = a + 0$.

Now s is a subgroup. Therefore $0 \in S$ or $0 \in -S$.

So $\quad a + 0 \in S + (-S).$

$\therefore \quad S \subseteq S + (-S).$

$\therefore \quad S = S + (-S).$

Also S must be closed with respect to multiplication.

$\therefore \quad a \in S, b \in S \Rightarrow ab \in S.$

Now ab is an arbitrary element of SS.

$\therefore \quad SS \subseteq S.$

The conditions are sufficient: Suppose S is a non-empty subset of R satisfying the two given conditions.

We have $SS \subseteq S \Rightarrow ab \in S \ \forall \ a, b \in S.$

Therefore S is closed with respect to multiplication.

Also $\quad S + (-S) = S \Rightarrow S + (-S) \subseteq S$

$\Rightarrow a + (-b) \in S$ if $a, b \in S$

$\Rightarrow$ S is a subgroup of the additive group of R.

$\therefore$ S is a subring of R.

Theorem 3:

An arbitrary intersection of subrings is a subring.

Proof:

Let R be a ring and let $\{S_t : t \in T\}$ be any family of subrings of R. Here T is an index set and is such that $\forall \ t \in T$, S_t is a subring of R.

Let $S = \bigcap_{t \in T} S_t = \{x \in R : x \in S_t \ \forall \ t \in T\}$

be the intersection of this family of subrings of R. Then to prove that S is also a subring of R.

Obviously $S \neq \varnothing$, since at least the zero element 0 of R is in $S_t \ \forall \ t \in T$.

Now let a, b be any two elements of S. Then

$a \in \bigcap_{t \in T} S_t \Rightarrow a \in S_t \ \forall \ t \in T$

and $b \in \bigcap_{t \in T} S_t \Rightarrow b \in S_t \ \forall \ t \in T$.

But $\forall \ t \in T$, S_t is a subring of R. Therefore

$$a, b \in S_t \Rightarrow a - b, ab \in S_t \ \forall \ t \in T.$$

Consequently $a - b, ab \in \bigcap_{t \in T} S_t$. Thus we have shown that

$$a, b \in \bigcap_{t \in T} S_t \Rightarrow a - b, ab \in \bigcap_{t \in T} S_t.$$

Therefore $\bigcap_{t \in T} S_t$ is a subring of R.

Intersection of Subrings

Theorem 4:

The intersection of two subrings is a subring.

Proof:

Let S_1 and S_2 be two subrings of a ring R. Then $S_1 \cap S_2$ is not empty since at least $0 \in S_1 \cap S_2$.

Now in order to prove that $S_1 \cap S_2$ is a subring, it is sufficient to prove that

1. $a \in S_1 \cap S_2, b \in S_1 \cap S_2 \Rightarrow S_2 \Rightarrow a - b \in S_1 \cap S_2$ and
2. $a \in S_1 \cap S_2, b \in S_1 \cap S_2 \Rightarrow S_2 \Rightarrow ab \in S_1 \cap S_2$.

We have $a \in S_1 \cap S_2 \Rightarrow a \in S_1, a \in S_2$

$b \in S_1 \cap S_2 \Rightarrow b \in S_1, b \in S_2$.

Now S_1 and S_2 are both subrings.

$\therefore$ $a \in S_1, b \in S_1 \Rightarrow a - b \in S_1$ and $ab \in S_1$

and $a \in S_2, b \in S_2 \Rightarrow a - b \in S_2$ and $ab \subset S_2$.

Now $a - b \in S_1, a - b \in S_2 \in a - b \in S_1 \cap S_2$

and $ab \in S_1, ab \in S_2 \Rightarrow ab \in S_1 \cap S_2$.

Thus $a \in S_1 \cap S_2, b \in S_1 \cap S_2$

$\Rightarrow a - b \in S_1 \cap S_2$ and $ab \in S_1 \cap S_2$.

$\therefore S_1 \cap S_2$ is a subring of R.

1.13 SMALLEST SUBRING CONTAINING A GIVEN SUBSET OF A RING

Suppose R is a ring and M is any subset of R. Further suppose that S is a subring of r such that $M \subseteq S$ and if T is any subring of R containing M then $S \subseteq T$. Then S is called the subring of R generated by the subset M. In short if S is the smallest subring of R containing M, then S is called the subring generated by M. Also we write S = (M).

Theorem:

The intersection of the family of subrings which contain a given subset M of a ring R is the smallest subring containing the subset M.

Proof:

The family of subrings which contain M is not empty since at least R is a subring of R which contains M. Further the intersection of all subrings of R which contain M is also a subring of R and M is contained in this intersection. Also this intersection will be contained in any subring of R which contains M. Therefore this intersection is the smallest subring of R containing M *i.e.*, it is the subring of R generated by M.

Example 1:

Show that the set of matrices $\begin{bmatrix} a & b \\ 0 & c \end{bmatrix}$ *is a subring of the 2 × 2 matrices with integral elements.*

Solution:

Let R be the ring of 2 × 2 matrices and let M be the subset of R and let the elements of M be matrices of the type

$$\begin{bmatrix} a & b \\ 0 & c \end{bmatrix}.$$

Let $A = \begin{bmatrix} a_1 & b_1 \\ 0 & c_1 \end{bmatrix}, B = \begin{bmatrix} a_2 & b_2 \\ 0 & c_2 \end{bmatrix}$ be any two elements of M.

Then $A - B = \begin{bmatrix} a_1 - a_2 & b_1 - b_2 \\ 0 & c_1 - c_2 \end{bmatrix}$ which is obviously an element of M.

Also $AB = \begin{bmatrix} a_1 & b_1 \\ 0 & c_1 \end{bmatrix} = \begin{bmatrix} a_2 & b_2 \\ 0 & c_2 \end{bmatrix} = \begin{bmatrix} a_1a_2 & a_1b_2 + b_1c_2 \\ 0 & c_1c_2 \end{bmatrix}$

which is obviously an element of M.

$\therefore$ M is a subring of R.

Example 2:

Let r be the ring of integers. Let m be any fixed integer and let S be any subset of R such that

S = {..., –3m, – 2m, – m, 0, 2m, m,...}. Then S is a subring of R.

Let a = rm and b = sm be any two elements of S. Then r and s are some integers. We have

$$a - b = rm - sm = (r - s)\, m \text{ and } ab = (rm)\,(sm) = (rsm)\, m.$$

Since r – s is some integer and ab = (rsm) is also some. Integer, therefore both a – b and ab are elements of S. Hence S is a subring of R.

Example 3:

The set of integers is a subring of the ring of rational numbers.

If a, b $\in$ I, then a – b $\in$ I and ab $\in$ I.

$\therefore$ I is a subring of the ring of rational numbers.

Example 4:

Let R be the ring of all 2 × 2 matrices over the field of real numbers. Let M be a subset of R and let the elements of M be matrices of the type

$$\begin{bmatrix} a & 0 \\ b & 0 \end{bmatrix}$$

i.e., matrices in which the elements of second column are all zeros. The M is a subring of R.

Let $A = \begin{bmatrix} a_1 & 0 \\ b_1 & 0 \end{bmatrix}$, $B = \begin{bmatrix} a_2 & 0 \\ b_2 & 0 \end{bmatrix}$ be any two elements of M.

Then

$$A - B = \begin{bmatrix} a_1 - a_2 & 0 \\ b_1 - b_2 & 0 \end{bmatrix}, \text{ Also } AB = \begin{bmatrix} a_1 & 0 \\ b_1 & 0 \end{bmatrix}, B = \begin{bmatrix} a_2 & 0 \\ b_2 & 0 \end{bmatrix} = \begin{bmatrix} a_1 - a_2 & 0 \\ b_1 - b_2 & 0 \end{bmatrix}.$$

Now A – B and AB are both members of M since the second column of A – B and also of AB consists of zeros only.

$\therefore$ M is subring of R.

1.14 SUBFIELDS

Definition:

Let F be a field. A non-empty subset K of the set F is said to be a subfield of F if K is closed with respect to the operations of addition and multiplication in F and K itself is a field for these operations.

Conditions for a Subfield

Theorem:

The necessary and sufficient conditions for a non-empty subset K of a field F to be a subfield of F are

1. $a \in K, b \in K \Rightarrow a - b \in K$,
2. $a \in K, 0 \neq b \in K \Rightarrow ab^{-1} \in K$.

Proof:

The conditions are necessary: Suppose K is a subfield of the field F. Now K is a group with respect to addition. Therefore $b \in K \Rightarrow -b \in K$. Also K is closed with a respect to addition.

$$\therefore \quad a \in K, b \in K \Rightarrow a + (-b) \in K \Rightarrow a - b \in K.$$

Now each non-zero element of K possesses multiplicative inverse. Therefore $0 \neq b \in K \Rightarrow b^{-1} \in K$.

But K is closed with respect to multiplication.

$$\therefore \quad a \in K, 0 \neq b \in K \Rightarrow ab^{-1} \in K.$$

Hence the conditions are necessary.

The conditions are sufficient: Suppose K is a non-empty subset of F and the conditions (1) and (2) are satisfied.

As we have proved in subrings, we can prove that with the help of condition (1), (K, +) is an abelian group. (Give the same proof here].

Now let a be any non-zero element of K. Then from (2) we have $a \in K$ $0 \neq a \in K \Rightarrow aa^{-1} \in K \Rightarrow 1 \in K$

Now $1 \in K$, therefore again from (2), we have

$$1 \in K \quad 0 \neq a \in K \Rightarrow 1a^{-1} \in K \Rightarrow a^{-1} \in K.$$

$\therefore$ Each non-zero element of K possesses multiplicative inverse.

Now let $a \in K$ and $0 \neq a \in K$. Then $b^{-1} \in K$.

From (2), we have

$$a \in K, 0 \neq b^{-1} \in K \Rightarrow a\,(b^{-1})^{-1} \in K \Rightarrow ab \in K$$

Also if b = 0, then ab = 0 and $0 \in K$.

$\therefore \quad ab \in K \; \forall \; a, b, \in K.$

Associativity of multiplication and distributivity of multiplication over addition must hold in K since they hold in F.

Example 1:

If R is a ring, show that

$$Z(R) = \{x \in R : xy = yx \;\; \forall y \in R\}$$

is a subring of R. Further show that Z(R) is a field if R is a division ring.

Solution:

We have $0y = 0 = y0 \;\; \forall y \in R$. Therefore $0 \in Z(R)$ and so $Z(R) \neq \varnothing$.

Now let $z_1, z_2 \in Z(R)$. Then

$$z_1y = yz_1 \text{ and } z_2y = yz_2 \;\; \forall y \in R.$$

Now $\forall y \in R$, we have

$$(z_1 - z_2)\, y = z_1y - z_2y = yz_1 - yz_2 = y\,(z_1 - z_2)$$

and $(z_1z_2)\, y = z_1\,(z_2y) = z_1\,(yz_2)\, z_2 = (z_1y)\, z_2 = (yz_1)\, z_2 = y\,(z_1z_2)$.

$\therefore$ by definition of Z(R), both $z_1 - z_2$ and $z_1z_2 \in Z(R)$.

Thus $z_1, z_2 \in Z(R) \Rightarrow z_1 - z_2 \in Z(R)$ and $z_1z_2 \in Z(R)$.

$\therefore \quad$ Z(R) is a subring of R.

Now suppose R is a division ring *i.e.*, R is a ring with unity and every non-zero element of R possesses multiplicative inverse. Then to prove that Z(R) is a field.

Z(R) is a commutative ring. Let $z_1, z_2 \in Z(R)$.

Now $z_1 \in Z(R) \Rightarrow z_1y = yz_1 \;\; \forall y \in R$. Since $z_2 \in R$, therefore $z_1z_2 = z_2z_1$.

Thus $z_1z_2 = z_2z_1 \;\; \forall z_1, z_2 \in Z(R)$ and so Z(R) is a commutative ring.

The Ring Z(R) Possesses Multiplicative Identity: If 1 denotes the unity element of the division ring R, then $1y = y = y1 \;\; \forall y \in R$. Therefore $1 \in Z(R)$ and is also the unity element of Z(R). Thus Z(R) is also a ring with unity.

Each Non-zero Element of the Ring Z(R) Possesses Multiplicative Inverse: Let $0 \neq z \in Z(R)$ and let z^{-1} denote the multiplicative inverse of z in the division ring R. We shall show that $z^{-1} \in Z(R)$.

We have $z \in Z(R) \Rightarrow zy = yz \;\; \forall y \in R$

$\Rightarrow \quad z^{-1}(zy)z^{-1} = z^{-1}(yz)z^{-1} \quad \forall y \in R$

$\Rightarrow \quad (z^{-1}z)yz^{-1} = (z^{-1}y)zz^{-1} \quad \forall y \in R$

$\Rightarrow \quad 1(yz^{-1}) = (z^{-1}y)1 \qquad \forall y \in R$

$\Rightarrow \quad yz^{-1} = z^{-1}y \quad \forall y \in R.$

$\therefore$ by the definition of Z(R), $z^{-1} \in Z(R)$.

Since $zz^{-1} = 1 = z^{-1}z$. where 1 is the unity element of the ring Z(R), therefore z^{-1} is also the multiplicative inverse of z in the ring Z(R).

Thus each non-zero element of the ring Z(R) is inversible.

Since Z(R) is a commutative ring with unity and each non-zero element of Z(R) is inversible, therefore Z(R) is a field.

Now $S_1 \cup S_2$ is not a subring of the ring of integers because $S_1 \cup S_2$ is not closed for addition.

We have $2 \in S_1 \cup S_2$ is because $2 \in S_1$ and $3 \in S_1 \cup S_2$ because $3 \in S_2$. But $2 + 3 = 5 \notin S_1 \cup S_2$ because neither $5 \in S_1$ nor $5 \in S_2$. Thus $S_1 \cup S_2$ is not closed for addition. Hence $S_1 \cup S_2$ is not a subring of the ring of integers.

However $S_1 \cup S_3 = S_1$ because $S_3 \subset S_1$. Thus $S_1 \cup S_2$ is a subring of the ring of integers.

From the above example it is obvious that the union of two subrings may or may not be a subring.

Example 2:

Show that the set of even integers forms a subring of the ring of integers.

Solution:

Let (I, +, .) be the ring of integers.

Let $S = 2I = \{..., -6, -4, -2, 0, 2, 4, 6, ...\}$ be the set of even integers. Then $S \subset I$.

To prove that S is a subring of the ring of integers.

Let $a = 2r$ and b 2s be any two elements of S where r and s are some integers.

Then $a - b = 2r - 2s = 2(r - s)$ which is an even integer and so $a - b \in S$.

Also $ab = (2r)(2s) = 2(2rs)$ which is also an even integer and so $ab \in S$.

Thus, a, b $\in$ S $\Rightarrow$ a – b $\in$ S and ab $\in$ S. Hence the set of even integers S is a subring of the ring of integers.

Example 3:

Prove or disprove that any subring of a non-commutative ring is non-commutative.

Solution:

A subring of a non-commutative ring may be commutative as is obvious from the following example.

Let M be the ring of all 2 × 2 matrices with elements as integers for addition and multiplication of matrices as the tow ring operations. Then M is a non-commutative ring because the operation of multiplication of matrices on the set M is not commutative.

Let S be the subset of M consisting of matrices of the type

$$\begin{bmatrix} a & 0 \\ 0 & 0 \end{bmatrix}, \text{ where a is any integer.}$$

Let $A = \begin{bmatrix} a_1 & 0 \\ 0 & 0 \end{bmatrix}, B = \begin{bmatrix} a_2 & 0 \\ 0 & 0 \end{bmatrix}$ be any two members of the set S.

The $A - B = \begin{bmatrix} a_1 - a_2 & 0 \\ 0 & 0 \end{bmatrix} \in S$

and $AB = \begin{bmatrix} a_1 a_2 & 0 \\ 0 & 0 \end{bmatrix} \in S$

$\therefore$ S is a subring of the ring M.

Now the subring S of the ring M is a commutative ring. For let

$$A = \begin{bmatrix} a_1 & 0 \\ 0 & 0 \end{bmatrix} \text{ and } B = \begin{bmatrix} a_2 & 0 \\ 0 & 0 \end{bmatrix}$$

be any two members of S. Then

$$AB = \begin{bmatrix} a_1 a_2 & 0 \\ 0 & 0 \end{bmatrix} \text{ and } BA = \begin{bmatrix} a_2 a_1 & 0 \\ 0 & 0 \end{bmatrix}.$$

Since $a_1 a_2 = a_2 a_1$, therefore AB = BA.

Thus, multiplication of matrices is a commutative operation on the set S. Hence the subring S of the non-commutative ring M is a commutative ring.

Example 4:

Show that the set of all 2-rowed matrices of the form

$$\begin{bmatrix} a & 0 \\ b & c \end{bmatrix}$$

where a, b, c are integers is a subring of the ring M of all 2-rowed matrices with integral entries.

Solution:

Let M be the set of all 2 × 2 matrices with elements as integers. Then M is a ring for addition and multiplication of matrices as the two ring operations.

Let S be the subset of M consisting of matrices of the type

$$\begin{bmatrix} a & 0 \\ b & c \end{bmatrix}, \text{ where a, b, c are integers.}$$

To prove that S is a subring of the ring M.

Let $A = \begin{bmatrix} a_1 & 0 \\ b_1 & c_1 \end{bmatrix}, B = \begin{bmatrix} a_2 & 0 \\ b_2 & c_2 \end{bmatrix}$ be any two members of the set S.

The $A - B = \begin{bmatrix} a_1 - a_2 & 0 \\ b_1 - b_2 & c_1 - c_2 \end{bmatrix}$ which is obviously a member of the set S.

Also $AB = \begin{bmatrix} a_1a_2 & 0 \\ b_1a_2 + c_1b_2 & c_1c_2 \end{bmatrix}$ which is also a member of the set S.

Thus A, B $\in$ S $\Rightarrow$ A – B $\in$ S and AB $\in$ S

Hence S is a subring of the ring M

1.15 CHARACTERISTIC OF A RING

Definition: *Let R be a ring with zero element 0 and suppose there exists a positive integer n such that na = a + a +... upto n terms = 0 for every a $\in$ R. The smallest such positive integer n is called the characteristic of the ring. If there exists no such positive integer, then R is said to be for characteristic zero or infinite.*

If any element of a ring R is of order zero when regarded as an element of the additive group (R, +), Then R will be of zero characteristic.

The ring of integers is of characteristic zero. The ring of rational numbers is also of characteristic zero.

If $I_6 = \{0, 1, 2, 3, 4, 5,\}$, then the ring $(I, +_6, \times_6)$ *i.e.*, the ring of integers modulo 6 has characteristic 6 since 6x = 0 for every x in the ring. Obviously, no integer smaller than 6 satisfies this property. For instance, 5 cannot be the characteristic, since 5(2) = 4 in I_6 and $4 \neq 0$.

Theorem 1:

The characteristic of an integral domain is 0 or n > 0 according as the order of any non-zero element regarded as a member of the additive group of the integral domain is either 0 or n.

Proof:

Let D be an integral domain.

If a non-zero element a of D is of order zero, then the characteristic of D is zero.

Let the order of the non-zero element a be finite and equal to n. Then na = 0.

Suppose b is any other non-zero element of D.

We have na = 0

$\Rightarrow$ (na) b = 0

$\Rightarrow$ (a + a + a + ... upto n terms) b = 0

$\Rightarrow$ (ab + ab + ab + ... upto n terms) = 0

$\Rightarrow$ a (b + b + b + ... upto n terms) = 0

$\Rightarrow$ a (nb) = 0.

But D is without zero divisors. Therefore $a \neq 0$ and $a(nb) = 0 \Rightarrow nb = 0$.

But the order of a is n $\Rightarrow$ n is the least positive integer such that na = 0. Also we have n0 = 0. Thus n is the least positive integer such that nx = 0 $\forall\, x \in D$, hence D is of characteristic n.

Theorem 2:

The characteristic of a ring with unity is 0 or n > 0 according as the unity element 1 regarded as a member of the additive group of the ring has the order zero or n.

Proof:

Let R be a ring with unity element 1. If 1 has order zero, then the characteristic of the ring is zero.

Suppose 1 is of finite order n so that

$$1 + 1 + 1 + \ldots \text{ upto n terms} = 0 \text{ i.e., } n1 = 0.$$

Let a be any element of R. Then, we have

$$\begin{aligned} na &= a + a + a + \ldots \text{ upto n terms} \\ &= 1a + 1a + 1a + \ldots \text{ upto n terms} \\ &= (1 + 1 + 1 + \ldots \text{ upto n terms})\, a && \text{[by dist. law]} \\ &= (n1)\, a = 0a = 0. \end{aligned}$$

$\therefore$ order of a is $\leq$ n.

Hence the characteristic of the ring is n.

Theorem 3:

Each non-zero element of an integral domain D, regarded as a member of the additive group of D, is of the same order.

Proof:

Let D be an integral domain, Suppose a is a non-zero element of D and o (a) is finite and say, equal to n.

Suppose b is any other non-zero element of D and o (b) = m.

We have $o(a) = n \Rightarrow na = 0$

$$\begin{aligned} &\Rightarrow nb = 0 \\ &\Rightarrow o(b) \leq n \Rightarrow m \leq n. \end{aligned}$$

Similarly $o(b) = m \Rightarrow mb = 0 \Rightarrow a(mb) = 0$

$$\begin{aligned} &\Rightarrow a(b + b + \ldots \text{ upto m times}) = 0 \\ &\Rightarrow (ab + ab + ab + \ldots \text{ upto m times}) = 0 \\ &\Rightarrow (a + a + a + \ldots \text{ upto m times})\, b = 0 \\ &\Rightarrow (ma)\, b = 0 \\ &\Rightarrow ma = 0 && [\because b \neq 0 \text{ and D is without zero divisors}] \\ &\Rightarrow o(a) \leq m \Rightarrow n \leq m. \end{aligned}$$

Now $m \leq n$, $n \leq m \Rightarrow m = n$. Hence $o(a) = o(b)$.

Also if o (a) is zero, then o (b) cannot be finite. Because o (b) = m $\Rightarrow$ ma = 0 i.e, the order of a is finite. Hence o (b) must also be zero. Hence the theorem.

Theorem 4:

The characteristic of an integral domain is either 0 or a prime number.

Proof:

Suppose D is an integral domain. Let $0 \neq a \in D$. If o (a) is zero, then the characteristic of D is 0. If o (a) is finite, let o (a) = p. Then the characteristic of D will be p. We are to prove that p must be prime.

Suppose p is not prime. Let $p = p_1p_2$ where $p_1 \neq 1$, $p_2 \neq 1$ and $p_1 < p$ also $p_2 < p$.

Since D is an integral domain, therefore the product to two non-zero elements of d cannot be equal to 0.

$\therefore \quad aa \neq 0$ *i.e.*, $a^2 \neq 0$.

Now in an integral domain two non-zero elements are of the same order.

$$\therefore \quad o(a) = p \Rightarrow o(a^2) = p \Rightarrow pa^2 = 0$$

$$\Rightarrow (p_1p_2)\, a^2 = 0 \qquad [\because \; p = [p_1p_2]$$

$$\Rightarrow (a^2 + a^2 + a^2 + \ldots \text{ upto } p_1p_2 \text{ terms}) = 0$$

$$\Rightarrow (p_1a)\,(p_1a) = 0$$

$\Rightarrow$ either $p_1a = 0$ or $p_2a = 0$ [$\because$ D is without zero divisors]

But $p_1 < p$ and $p_2 < p$. Also p is the least positive integer such that pa = 0. Hence p must be prime.

1.16 CHARACTERISTIC OF FIELD

Every field is an integral domain. Therefore the characteristic of a field F is 0 or n > 0 according as any non-zero element (in particular the unit element 1) of F is of order 0 or n.

Thus in order to find the characteristic of a field F, we should find the order of the unit element 1 of F when regarded as a member of the additive group of F. If the order of 1 is zero then f is of characteristic 0. If the order of 1 is finite, say, n then the characteristic of F is n.

Then characteristic of the field of real numbers is 0.

The characteristic of the finite field $(I_7, +_7, \times_7)$ is 7 where $I_7 = \{0, 1, 2, 3, 4, 5, 6\}$.

1.17 ORDERED INTEGRAL DOMAINS

Definition: *An integral domain (D, +, ·) is said to be ordered if D contains a subset D_+ such that*

1. *D_+ is closed with respect to addition and multiplication as defined on D.*
2. *$\forall a \in D$, one and only one of $a = 0$, $a \in D_+$,... $a \in D_+$ holds (principle of Trichotomy).*

The elements of D_+ are called the positive elements of D, all other non-zero elements of D are called negative elements of D.

Ordered Field: *A field (F, +, ·) is to be ordered if it is ordered as an integral domain.*

The integral domain $(I_+, +, \cdot)$ of all integers is ordered.

The set I_+ of all positive integers is the set of the positive elements of this integral domain. We know that the sum and product of two positive integers is again a positive integer *i.e.*, I_+ is closed with respect to addition and multiplication. If $a \in I$, then either is a zero or positive or negative *i.e.*, either $a = 0$ or $a \in I_+$ or $-a \in I_+$.

The field of rational numbers is an ordered field. The field of real numbers is also an ordered field. But the field of complex numbers is not an ordered field.

Theorem 1:

The field (C, +, ·) of complex numbers is not ordered.

Proof:

Suppose C is an ordered field and C_+ is the set of positive elements of this field. The additive identity *i.e.*, the zero element is $0 + i0$.

Now $\quad i \neq 0$.

By the principle of trichotomy either $i \in D_+$ or $-i \in C_+$.

Now C_+ is closed with respect to multiplication.

$\therefore i \in C_+, -1 \in C_+ \Rightarrow i(-1) = -i \in C_+$. Thus if $i \in C_+$, its additive inverse $-i$ also belongs to C_+. This contradicts the principle of trichotomy *i.e.*, the condition (2) of the definition of an ordered integral domain.

Similarly if we assume that $-1 \in C_+$, we can show that its additive inverse i also belongs to C_+. This again contradicts the principle of trichotomy.

Hence the field of complex numbers is not an ordered field.

Theorem 2:

The field $(I_p, +_p, \times_p)$ where p is a prime and $I_p = \{0, 1, ..., p-1\}$ is not ordered.

Proof:

Suppose I_p is an ordered field and P is the set of positive elements of the field. The zero element of this field is 0.

Now $1 \neq 0$. The additive inverse of 1 is $p-1$. According to the definition of ordered field either $1 \in P$ or its additive inverse $p - 1 \in P$.

But P is closed with respect to $+_p$.

Therefore $1 \in P \Rightarrow 1 +_p 1 +_p 1 +_p$ ...upto $p - 1$ times $\in P$

$$\Rightarrow p - 1 \in P.$$

This contradicts the principle of trichotomy. Similarly assuming that $p - 1 \in P$ we can show that its additive inverse 1 also belongs to P. This again contradicts the principle of trichotomy. Hence the given field is not an ordered field.

Theorem 3:

Let D be an integral domain with unity element 1. If D is an ordered integral domain show that 1 is a positive element of D

Proof:

Let D be an ordered integral domain with unity element 1. Let D_+ denote the set of positive elements of D.

Suppose $1 \notin D_+$.

Now $1 \neq 0$. Since $1 \notin D_+$, therefore by the definition of an ordered integral domain,

$$-1 \in D_+$$

$\Rightarrow (-1)(-1) \in D_+$ [$\because$ D_+ is closed with respect to multiplication]

$\Rightarrow 1 \in D_+$ which is a contradiction.

Hence $1 \in D_+$ *i.e.*, 1 is a positive element of D.

Order Relations in an Ordered Integral Domain

Definition: *Let D be an ordered integral domain and D_+ be the set of positive elements of D. Then we define less than' (<) 'greater than' (>) relations in D as follows:*

For all a, b, $\in$ *D, we have*

1. $a > b$ when $a - b \in D_+$,
2. $a < b$ when $b - a \in D_+$.

Obviously a > iff b < a.

Theorem:

The order relation in an ordered integral domain is transitive i.e., $a > b,\ b > c \Rightarrow a > c$.

Proof:

Let D be an ordered integral domain and let D_+ be the set of positive elements of D.

We have $a > b \Rightarrow a - b \in D_+$ [by def. of >]

and $b > c \Rightarrow b - c \in D_+$.

Now D_+ is closed with respect to addition.

$$\therefore a - b \in D_+,\ b - c \in D_+ \Rightarrow (a - b) + (b - c) \in D_+$$

$$\Rightarrow a - c \in D_+ \Rightarrow a > c.$$

Example 1:

Show that in an integral domain all non-zero elements generate additive cholic groups of the same order which is equal to the characteristic of the integral domain.

Solution:

Let d be an integral domain.

First we shall prove that each non-zero element of D, regarded as a member of the additive group of D, is of the same order.

Now we know that the order of a cyclic group is equal to the order of its generator. Hence in an integral domain all non-zero elements generate additive cyclic groups of the same order.

In the end we shall prove that the characteristic of the integral domain D is equal to the order of any non-zero element of D regarded as a member of the additive group of D.

Example 2:

Give without proof, an example of an integral domain which contains only five elements. Is this an ordered integral domain? Give reason.

Solution:

The integral domain $(\{0, 1, 2, 3, 4\}, +_5, \times_5\}, +_5, \times_5)$ contains only five elements

This integral domain is not an ordered integral domain as shown below.

Suppose the above integral domain is an ordered integral domain and P is the set of positive elements of this integral domain. The zero element of this integral domain is 0.

Now 1 is an element of this integral domain and $1 \neq 0$. The additive inverse of 1 is 4 because $1 +_5 4 = 0 = 4 +_5 1$.

According to the definition of an ordered integral domain either $1 \in P$ or its additive inverse $4 \in P$.

But P is closed with respect to $+_5$.

$\therefore 1 \in P \Rightarrow 1 +_5 1 +_5 1 +_5 1 \in P \Rightarrow 4 \in P$.

This contradicts the principle of trichotomy.

Similarly $4 \in P \Rightarrow 4 +_5 4 +_5 4 +_5 4 \in P \Rightarrow 1 \in P$. This again contradicts the principle of trichotomy.

Hence, the integral domain $(\{0, 1, 2, 3, 4\}, + 5, \times 5)$ is not an ordered integral domain.

Example 3:

Define the characteristic of a ring and prove that if R is a finite ring then the characteristic of R is finite and $\neq 0$.

Solution:

Let $(R, +, \cdot)$ be a finite ring having n elements.

From our study of group theory we know that if a is any element of a finite group G of order n, then $a^n = e$, where e is the identity of the group.

Now the identity of the additive group $(R, +)$ of the ring $(R, +, \cdot)$ is 0. Therefore if a sin ay element of R, then we have $na = 0$.

Thus we have $na = 0 \ \forall \ a \in R$.

Therefore the characteristic of the ring R is $\leq n$ and cannot be infinite or zero.

Hence if R is a finite ring then the characteristic of R is finite and $\neq 0$.

Example 4:

Let x, y be commutative elements of a ring R of characteristic two. Show that $$(x + y)^2 = x^2 + y^2 = (x - y)^2$$

Solution:

Since the ring R is of characteristic two, therefore

$$a + a = 0 \ \forall \ a \in R.$$

Now let x, y $\in$ R and xy = yx.

Then $(x + y)^2 = (x + y)(x + y)$

$= x^2 + xy + yx + y^2$, by dist, laws

$= x^2 + xy + xy + y^2$ $[\because \ xy = yx]$

$= x^2 + 0 + y^2$ [$\because$ xy $\in$ R and R is of characteristic two implies xy + xy = 0]

$= x^2 + y^2$.

Again $(x - y)^2 = (x - y)(x - y) = x^2 - xy - yx + y^2$

$= x^2 - xy - xy + y^2 = x^2 - (xy + xy) + y^2$

$= x^2 - 0 + y^2 = x^2 + y^2$.

Example 5:

Let R be an non-zero ring such that for all $a \in R$, $a^2 = a$. Prove, that R is a commutative ring of characteristic 2.

Solution:

Do your self.

Hint: We have proved there in part (1) of the question that

$$a + a = 0 \ \forall \ a \in R.$$

Therefore the ring R is of characteristic two.

Again in parts (2) and (3) of that question we have proved that the ring R is commutative.

Ordered Field

Definition: A field (F, + , ·) is said to be ordered if F contains a subset F_+ such that

1. F_+ is closed with respect to addition and multiplication as defined on F.
2. $\forall$ a $\in$ F, one and only on of a = 0, a $\in F_+$, – a $\in F_+$ holds *(Principle of Trichotomy).*

The elements of F_+ are called the positive elements of F, all other non-zero elements of F are called negative elements of F.

An Example of an Ordered Field: The field of real numbers (R, +, ·) is an ordered field. The set R_+ of all positive real numbers is the set of the positive elements of this field. We know that the sum and product of two positive real numbers is again a positive real number *i.e.*, R_+ is closed with respect to addition and multiplication. If a ∈ R, then either a is zero or positive or negative *i.e.*, either a = 0 or a ∈ R_+ or –a ∈ R_+.

The field of rational numbers (Q, +, ·) is also an ordered field. But the field of complex numbers (C, + , ·) is not an ordered field.

Example:

Show that every finite integral domain is of finite characteristic.

Solution:

Let (R, +, ·) be a finite integral domain.

Let a be any non-zero element of R. Then we know that the characteristic of the integral domain (R, +, ·) is equal to the order of the element a regarded as a member of the additive group of the integral domain *i.e.*, regarded as a member of the additive group (R, +).

But from our study of the order of an element in the group theory we know that the order of every element of a finite group is finite. Therefore the order of a as a member of the additive group (R, +) is finite.

Hence very finite integral domain is of finite characteristic.

1.18 IDEALS

Definition:

(a) Left Ideal: *A non-empty subset S of a ring R is said to be a left ideal of R if:*

1. *S in a subgroup of R with respect to addition.*
2. *rs ∈ S ∀ r ∈ R and ∀ s ⊂ S.*

(b) Right Ideal: *A non-empty subset S of a ring R is said to be a right ideal of R if:*

1. *s is a subgroup of R under addition,*
2. *sr ∈ R ∀ r ∈ R and ∀ s ∈ R*

(c) Ideal: A non-empty subset S of a ring R is said to be an *ideal* (also a *two sided ideal*) if and only if it is both a left ideal and a right ideal. *Thus, f a non-empty subset S of a ring R is said to be an ideal of R if:*

1. *S is a subgroup of R under addition i.e., S is a subgroup of the additive group of R.*
2. *rs ∈ S and sr ∈ S for every r ∈ R and for every s ∈ S.*

If S is an ideal of a ring R, then S is also a subring of R. The obvious reason is that S is a subgroup of R under addition and from condition (2), we have xs ∈ S ∀ x s ∈ S because x ∈ S ⇒ x ∈ R. Thus s is closed with respect to multiplication. Therefore S is a subring of *R. Thus every ideal of a ring R is also a subring of R.* But every subring is not an ideal. An ideal requires a stronger closure property than the subring. If S is a subgroup of R under addition, then S will be a subring if S is closed with respect to multiplication *i.e.*, the product of two elements of S with any element of R is in S.

If R is commutative ring, then very left ideal will also be a right ideal. Therefore in a commutative ring every left (right) ideal is an ideal.

Notes:

1. If we are to prove that a non-empty subset S of a ring R is an ideal of R, then it is sufficient to prove that

   (i) a ∈ S, b ∈ S ⇒ a – b ∈ S, and

   (ii) rs ∈ S and sr ∈ S ∀ r ∈ R and ∀ r ∈ S.

   Obviously (i) is a sufficient condition for S to be a subgroup of R under addition.

   If R is a commutative ring, then the condition (ii) will be come more simple. Then it will become

   rs ∈ S ∀ r ∈ R and ∀ s ∈ S.

2. Every ring R always possesses two improper ideals: one R itself and the other consisting of 0 only. These are respectively know as the unit ideal and the null ideal.

   Any other ideals of R are called proper ideals. A ring having no proper ideals is called a simple ring.

Theorem 1:

An arbitrary intersection of left ideals of a ring is a left ideal of the ring.

Proof:

Let R be a ring and let $\{S_t : t \in T\}$ be any family of left ideals of R. Here T is an index set and is such that ∀ t ∈ T, S_t is a left ideal of R. Let $S = \bigcap_{t \in T} S_t = \{x \in R : x \in S_t \ \forall\, t \in T\}$

be the intersection of this family of left ideals of R. Then to prove that S is also a left ideal of R.

Obviously $S \neq \varnothing$, since at least 0 is in $S_t \ \forall \ t \in T$.

Now let a, b be any two elements of S. Then

$$a, b, \in S \Rightarrow a, b, \in S_t \ \forall \ t \in T$$

$$\Rightarrow a - b \in S_t \ \forall \ t \in T$$

$$[\because \forall \ t \in T, S_t \text{ is a left ideal of R}]$$

$$\Rightarrow a - b \in \bigcap_{t \in T} S_t \Rightarrow a - b \in S.$$

Now let a be any element of S and R be any element of R.

We have $a \ a \in S \Rightarrow a \ a \in \bigcap_{t \in T} S_t \ \forall \ t \in T$

$\Rightarrow \quad rs \in S_t \ \forall \ t \in T \quad [\because \forall \ t \in T, S_t \text{ is a left ideal of R}]$

$\Rightarrow \quad rs \in \bigcap_{t \in T} S_t \Rightarrow ra \in S.$

Thus $a, b, \in S \Rightarrow a - b \in S$ and $r \in R, a \in S \Rightarrow ra \in S$.

$\therefore$ S is a left ideal of R.

Theorem 2:

The intersection of any two left ideals of a ring is again a left ideal of the ring.

Proof:

Let I_1 and I_2 be two left ideals of a ring R. Then I_1, I_2 are subgroups of R under addition. Therefore $I_1 \cap I_2$ is also a subgroup of R under addition.

Now to show that $I_1 \cap I_2$ is a left ideal of R we are only to show that $r \subset R, s \in I_1 \cap I_2 \Rightarrow rs \in I_1 \cap I_2$.

We have $s \in I_1 \cap I_2 \Rightarrow s \in I_1, s \in I_2$

But I_1 and I_2 are left ideals of R. Therefore

$r \in R, s \in I_1 \Rightarrow rs \in I_1$ and $r \in R, s \in I_2 \Rightarrow \in I_2$.

Now $rs \in I_1, rs \in I_2 \Rightarrow rs \in I_1 \cap I_2$.

$\therefore I_1 \cap I_2$ is also a left ideal of R.

Note: A similar result can be proved for right ideals as well as for ideals.

1.19 SMALLEST LEFT IDEAL CONTAINING A GIVEN SUBSET

Definition: *Let M be a non-empty subset of a ring. Then a left ideal I of R is called the smallest left ideal of R containing M, if I contains M and if 1 is contained in every left ideal of R containing M.*

The smallest left ideal of R containing M is called the left ideal generated by M and will be denoted by (M)

It can be easily seen that the intersection of the family of left ideals containing M is the left ideal generated by M.

Remark: *A similar definition can be given for the right ideal generated by M as well as for the ideal generated by M. For this purpose simply replace the word 'left ideal' by 'right ideal' or by 'ideal'.*

Sum of Two Left Ideals

Theorem:

The left ideal generated by the union $I_1 \cup I_2$ of the left ideals is the set $I_1 + I_2$ consisting of the elements of R obtained on adding any element of I_1 to any element of I_2.

Proof:

Let $a_1 + a_2, b_1 + b_2 \in I_1 + I_2$.

Then $a_1, b_1 \in I_1$ and $a_2, b_2 \in I_2$.

Since I_1, I_2 are left ideals of R, therefore they are subgroups of the additive group of R. Therefore

$a_1, b_1 \in I_1 \Rightarrow a_1 - b_1 \in I_1$ and $a_2, b_2 \in I_2 \Rightarrow a_2\ b_2 \in I_2$.

Consequently $(a_1 + a_2) - (b_1 + b_2) = (a_1 - b_1) + (a_2 - b_2) \in I_1 + I_2$.

Therefore $I_1 + I_2$ is a subgroup of the additive group of R.

Now let $r \in R$ and $a_1 + a_2 \in I_1 + I_2$. Then $a_1 \in I_1, a_2 \in I_2$.

We have $r(a_1 + a_2) = ra_1 + ra_2 \in I_1 + I_2$.

[$\because I_1$ is a left ideal implies $ra_1 \in I_1$ and similarly $ra_2 \in I_2$]

$\therefore I_1 + I_2$ is a left ideal of R.

Since $0 \in I_2$, therefore $a_1 \in I_1$ can be written as $a_1 + 0$. Thus

$$a_1 \in I_1 \Rightarrow \in I_1 + I_2.$$

$$\therefore \quad I_1 \subseteq I_1 + I_2.$$

Similarly $I_2 \subseteq I_1 + I_2$.

$$\therefore \quad I_1 \cup I_2 \subseteq I_1 + I_2.$$

Thus $I_1 + I_2$ is a left ideal containing $I_1 \cup I_2$.

Also if any left ideal contains $I_1 \cup I_2$, then it must contain $I_1 + I_2$.

$\therefore I_1 + I_2$ is the smallest left ideal containing $I_1 \cup I_2$.

$\therefore I_1 + I_2$ = the left ideal generated by $I_1 \cup I_2 = (I_1 \cup I_2)$.

Note: A similar result can be proved for right ideals as well as for ideals.

Example 1:

Show that for a field F, the set of all matrices of the form

$$\begin{bmatrix} a & b \\ 0 & 0 \end{bmatrix}$$

For a, b, $\in$ F is a right ideal but not a lift ideal of the ring of all 2×2 matrices over the field F.

Solution:

Let R be the ring of all 2×2 matrices with elements in the field F for addition and multiplication of matrices as the two ring operations. Let M be the subset of R consisting of matrices of the form

$$\begin{bmatrix} a & b \\ 0 & 0 \end{bmatrix}, \text{ a, b, } \in \text{F}.$$

First to show that M is a right ideal of the ring R.

Let $\quad A = \begin{bmatrix} a_1 & b_1 \\ 0 & 0 \end{bmatrix}$ and $B = \begin{bmatrix} a_2 & b_2 \\ 0 & 0 \end{bmatrix}$ be any two elements of M.

Then $A - B = \begin{bmatrix} a_1 - a_2 & b_1 - b_2 \\ 0 & 0 \end{bmatrix} \in M.$

$\therefore$ M is a subgroup of the additive group of the ring R.

Now let $U = \begin{bmatrix} w & x \\ y & z \end{bmatrix}$ be any element of R and $A = \begin{bmatrix} a & b \\ 0 & 0 \end{bmatrix}$ be any element of M.

Then $AU = \begin{bmatrix} a & b \\ 0 & 0 \end{bmatrix} \begin{bmatrix} w & x \\ y & z \end{bmatrix} = \begin{bmatrix} aw + by & ax + bz \\ 0 & 0 \end{bmatrix} \in M$

Therefore M is a right ideal of the ring R.

But M is not a left ideal of R, since

$$\begin{bmatrix} 0 & 0 \\ 1 & 0 \end{bmatrix} \in R, \begin{bmatrix} 1 & 1 \\ 0 & 0 \end{bmatrix} \in M,$$

and the product $\begin{bmatrix} 0 & 0 \\ 1 & 0 \end{bmatrix}\begin{bmatrix} 1 & 1 \\ 0 & 0 \end{bmatrix} = \begin{bmatrix} 0 & 0 \\ 1 & 1 \end{bmatrix} \notin M.$

Hence M is not a left ideal of R,

Example 2:

If R is ring and $a \in R$ let $T = \{x \in R : ax = 0\}$. Prove that T is a right ideal of R.

Solution:

First we see that U is not empty because

$0 \in R$ is such that $a0 = 0$.

Let x_1, x_2 be any two elements of T. Then $ax_1 = 0$, $ax_2 = 0$.

We have $a(x_1 - x_2) = ax_1 - ax_2 = 0 - 0 = 0$.

$\therefore x_1 - x_2 \in T$.

$\therefore$ T is a subgroup of R under addition.

Now to show that T is a right ideal of R we are to show that $x \in T$, $y \in R \Rightarrow xy \in T$. But $x \in T \Rightarrow ax = 0$. If we show that $a(xy) = 0$, then xy will be an element of T.

We have $a(xy) = (ax)y = 0y = 0$.

$\therefore \quad xy \in T$.

$\therefore$ T is a right ideal of R.

Example 3:

If m is a fixed integer, the set P of integers given by,

$$P = \{xm : x \text{ is an integer}\}$$

is an ideal of the ring R of all integers.

Solution:

Let $x_1m - x_2m$ be any two elements of P. Then x_1 and x_2 are some integers.

We have $x_1m - x_2m = (x_1 - x_2)m \in P$ since $x_1 - x_2$ is also an integer.

$\therefore$ P is a subgroup of R under addition.

Now let r be any integer *i.e.*, r be any element of R and xm be any element of P. Then r (xm) = (rn) m ∈ P since rx is also an integer. Therefore P is a left ideal of R. But R is a commutative ring. Hence P is an ideal of R.

Example 4:

Distinguish between subrings and ideals in a ring. Show that the 2-rowed matrices of the form

$$\begin{bmatrix} a & 0 \\ b & c \end{bmatrix}$$

where a, b, c are integers form a subring of the ring of all 2-rowed matrices with integral entries. Is this subring an integral domain?

Solution:

Do yourself.

Example 5:

The set N of all 2 × 2 matrices of the form

$$\begin{bmatrix} a & 0 \\ b & 0 \end{bmatrix}$$

for a, b integers is a left ideal but not a right ideal in the ring R of all 2 × 2 matrices with elements as integers. Here N is the subset of R consisting of those elements whose second column contains only zeros.

Solution:

Let $A = \begin{bmatrix} a & 0 \\ b & 0 \end{bmatrix}, B = \begin{bmatrix} c & 0 \\ d & 0 \end{bmatrix}$ be any tow elements of N.

Then $A - B = \begin{bmatrix} a & 0 \\ b & 0 \end{bmatrix} - \begin{bmatrix} c & 0 \\ d & 0 \end{bmatrix} = \begin{bmatrix} a-c & 0 \\ b-d & 0 \end{bmatrix} \in N.$

∴ N is a subgroup of R under addition.

Now let $U = \begin{bmatrix} w & x \\ y & z \end{bmatrix}$ be any element of R and $A = \begin{bmatrix} a & 0 \\ b & 0 \end{bmatrix}$ be any element of N.

Then $UA = \begin{bmatrix} w & x \\ y & z \end{bmatrix}\begin{bmatrix} a & 0 \\ b & 0 \end{bmatrix} = \begin{bmatrix} wa+xb & 0 \\ ya+zb & 0 \end{bmatrix} \in N.$

Therefore N is a left ideal of R. It is not a right ideal, since

$$\begin{bmatrix} 1 & 0 \\ 1 & 0 \end{bmatrix} \in N, \begin{bmatrix} 1 & 2 \\ 0 & 1 \end{bmatrix} \in R,$$

and the product $\begin{bmatrix} 1 & 0 \\ 1 & 0 \end{bmatrix}\begin{bmatrix} 1 & 2 \\ 0 & 1 \end{bmatrix} = \begin{bmatrix} 1 & 2 \\ 1 & 2 \end{bmatrix}$ which is not an element of N.

Distinction Between Subrings and Ideals in a Ring: Let R be any ring and S be any non-empty subset of R.

So far as the operation of addition is concerned, whether S is a subring or an ideal of R, S must be a subgroup of the additive group of R *i.e.*, $a \in S, b \in S \Rightarrow a - b \in S$.

But so far as the operation of multiplication is concerned, for S to be a subring of R, the product of any two elements of S must remain in S *i.e.*, $a \in S, b \in S \Rightarrow ab \in S$. While on the other hand for S to be an ideal of R the product of any element of R and any element of S must remain in S *i.e.*, $r \in r, s \in S \Rightarrow rs \in S$ and $sr \in S$. Thus an ideal requires a stronger closure property for multiplication then a subring. *In fact every ideal is a subring while a subring may or may not be an ideal.* For example the set of integers is only a subring but not an ideal of the ring of rational numbers. On the other hand the set of even integers is a subring as well as an ideal of the ring of integers.

Second Part of the Question: Let M be the ring of all 2×2 matrices with elements as integers for addition and multiplication of matrices as the two ring operations.

Let S be the subset of M consisting of matrices of the form

$\begin{bmatrix} a & 0 \\ b & c \end{bmatrix}$, where a, b, c are any integers.

To show that S is a subring of M.

Let $A = \begin{bmatrix} a_1 & 0 \\ b_1 & c_1 \end{bmatrix}, B \begin{bmatrix} a_2 & 0 \\ b_2 & c_2 \end{bmatrix}$ be any two elements of S.

Then $A - B = \begin{bmatrix} a_1 - a_2 & 0 \\ b_1 - b_2 & c_1 - c_2 \end{bmatrix}$ which is obviously an element of S.

Also $AB = \begin{bmatrix} a_1 - a_2 & 0 \\ b_1 a_2 + c_1 b_2 & c_1 c_2 \end{bmatrix}$ which is also an element of S.

Hence S is a subring of the ring M.

The subring S of M is not an integral domain because it possesses zero divisors. The zero element of the subring S is the mull matrix $\begin{bmatrix} 0 & 0 \\ 0 & 0 \end{bmatrix}$.

Now $A = \begin{bmatrix} 1 & 0 \\ 0 & 0 \end{bmatrix}$ and $B = \begin{bmatrix} 0 & 0 \\ 1 & 0 \end{bmatrix}$ are two non-zero elements of S but

$AB = \begin{bmatrix} 0 & 0 \\ 0 & 0 \end{bmatrix}$ = the zero element of S. Thus both A and B are zero divisors.

Hence S is not an integral domain.

Example 6:

Consider the ring R of all 3 × 3 matrices of the type

$$\begin{bmatrix} a & b & c \\ 0 & d & e \\ 0 & 0 & f \end{bmatrix},$$

a, b, c, d, e, f are real numbers. Show that the set I of all matrices of the form

$$\begin{bmatrix} a & 0 & 0 \\ 0 & 0 & 0 \\ 0 & 0 & 0 \end{bmatrix}$$

is a left ideal of R, which is not a right ideal.

Solution:

First to show that I is a left ideal of R.

Let $A - \begin{bmatrix} a_1 & 0 \\ 0 & 0 \end{bmatrix}$ and $B = \begin{bmatrix} a_2 & 0 \\ 0 & 0 \end{bmatrix}$ be nay two elements of I.

Then $A - B = \begin{bmatrix} a_1 - a_2 & 0 \\ 0 & 0 \end{bmatrix} \in I.$

$\therefore$ I is a subgroup of the additive group of the ring R.

Now let $U = \begin{bmatrix} a & b & c \\ 0 & d & e \\ 0 & 0 & f \end{bmatrix}$ be any element of R and $A = \begin{bmatrix} p & 0 & 0 \\ 0 & 0 & 0 \\ 0 & 0 & 0 \end{bmatrix}$ be any element of I.

Then $UA = \begin{bmatrix} a & b & c \\ 0 & d & e \\ 0 & 0 & f \end{bmatrix} \begin{bmatrix} p & 0 & 0 \\ 0 & 0 & 0 \\ 0 & 0 & 0 \end{bmatrix} = \begin{bmatrix} ap & 0 & 0 \\ 0 & 0 & 0 \\ 0 & 0 & 0 \end{bmatrix} \in I$

Therefore I is a left ideal of the ring R.

But I is not a right ideal of R, since

$$\begin{bmatrix} 1 & 1 & 1 \\ 0 & 1 & 2 \\ 0 & 0 & 1 \end{bmatrix} \in R, \begin{bmatrix} 1 & 0 & 0 \\ 0 & 0 & 0 \\ 0 & 0 & 0 \end{bmatrix} \in I$$

and the product $\begin{bmatrix} 1 & 0 & 0 \\ 0 & 0 & 0 \\ 0 & 0 & 0 \end{bmatrix} \begin{bmatrix} 1 & 1 & 1 \\ 0 & 1 & 2 \\ 0 & 0 & 1 \end{bmatrix} = \begin{bmatrix} 1 & 1 & 1 \\ 0 & 0 & 0 \\ 0 & 0 & 0 \end{bmatrix} \notin I$

Hence I is not a right ideal or R.

Example 7:

Prove that the intersection of two ideals of a ring R is an ideal of R.

Solution:

Let s and T be two ideals of a ring R. Then S, T are subgroups of R under addition. Therefore $S \cap T$ is also a subgroup of R under addition.

Now to show that $S \cap T$ is a ideal of R, we are only to show that

$$r \in R, s \in S \cap T \Rightarrow rs \in S \cap T, sr \in S \cap T.$$

We have $s \in S \cap T \Rightarrow s \in S, s \in T$.

But S and T are ideals of R. Therefore

$$r \in R, s \in S \Rightarrow rs \in S, sr \in S$$

and $\quad r \in R, s \in T \Rightarrow rs \in T, sr \in T.$

Now $\quad rs \in S, rs \in T \Rightarrow rs \in S \cap T$

and $\quad sr \in S,\ sr \in T \Rightarrow sr \in S \cap T.$

$\therefore \quad S \cap T$ is also an ideal of R.

Example 8:

Show that the set M of all 2 × 2 matrices of the form

$$\begin{bmatrix} 0 & a \\ 0 & b \end{bmatrix}$$

a, b, integers is a left ideal but not a right ideal in the ring of all 2×2 matrices with elements as integers.

Solution:

Let R be the ring of all 2 × 2 matrices with elements as integers for addition and multiplication of matrices as the two ring operations. Let M be the subset of R consisting of matrices of the form

$\begin{bmatrix} 0 & a \\ 0 & b \end{bmatrix}$, where a, b are any integers. First to show that M is a left ideal of the ring R.

Let $\quad A = \begin{bmatrix} 0 & a_1 \\ 0 & b_1 \end{bmatrix}$ and $B = \begin{bmatrix} 0 & a_2 \\ 0 & b_2 \end{bmatrix}$ be any two elements of M.

Then $\quad A - B = \begin{bmatrix} 0 & a_1 - a_2 \\ 0 & b_1 - b_2 \end{bmatrix} \in M.$

$\therefore$ M is a subgroup of the additive group of the ring R.

Now let $U = \begin{bmatrix} w & x \\ y & z \end{bmatrix}$ be any element of R and $A = \begin{bmatrix} 0 & a \\ 0 & b \end{bmatrix}$ be any element of M.

Then $UA = \begin{bmatrix} w & x \\ y & z \end{bmatrix}\begin{bmatrix} 0 & a \\ 0 & b \end{bmatrix} = \begin{bmatrix} 0 & wa+xb \\ 0 & ya+zb \end{bmatrix} \in M$

Therefore M is a left ideal of R.

But M is not a right ideal of R. Since

$$\begin{bmatrix} 0 & 1 \\ 0 & 1 \end{bmatrix} \in M, \begin{bmatrix} 2 & 4 \\ 3 & 5 \end{bmatrix} \in R,$$

and the product $\begin{bmatrix} 0 & 1 \\ 0 & 1 \end{bmatrix}\begin{bmatrix} 2 & 4 \\ 3 & 5 \end{bmatrix} = \begin{bmatrix} 3 & 5 \\ 3 & 5 \end{bmatrix}$ which is not an element of M. Hence M is not a right ideal of R.

Example 9:

If U is an ideal of a ring R with unity and $1 \in U$ prove that $U = R$.

Solution:

We have $U \subseteq R$ since U is an ideal of R. Let x be any element of R. Since u is an ideal of R, therefore

$$1 \in U, x \in R \Rightarrow 1x \in U \Rightarrow x \in U.$$

$$\therefore \quad R \subseteq U.$$

$$\therefore \quad U = R.$$

Example 10:

Show that S is an ideal of S + T where S is any ideal of ring R, and T any subring of R.

Solution:

Since S is an ideal of R therefore S is subring of R. Also Ti is a subring of R. First we shall show that S + T is a subring of R. Let $a + \alpha$, $b + \beta \in S + T$, where a, b, $\in$ S and α, β, □$\in$ T.

Since S is a subring, therefore $a - b \in S$. Similarly $\alpha - \beta \in T$.

$\therefore (a + \alpha) - (b + \beta) = (a - b) + (\alpha - \beta) \in S + T.$

Also $(a + \alpha)(b + \beta) = ab + a\beta + \alpha b + \alpha\beta = (ab + a\beta + \alpha b) + \alpha\beta.$

Now S is subring. Therefore a, b $\in S \Rightarrow ab \in S.$

Also S is an ideal, therefore a, b, $\in$ S and $\alpha, \beta \in R \Rightarrow a\beta, \alpha b \in S.$ Therefore $ab + a\beta + \alpha b \in S.$

Further t is a subring implies $\alpha\beta \in T$ if $\alpha, \beta \in T.$

$\therefore (a + \alpha)(b + \beta) = (ab + a\beta + \alpha b) + \alpha\beta \in S + T.$

$\therefore$ S + T is a subring of R.

Since $0 \in T$, therefore $a \in S$ can be written as

$$a = a + 0 \in S + T.$$

$$\therefore \quad S \subseteq S + T.$$

Thus, $S \subseteq S + T$ and S + T is a subring of R. Since S is an ideal of R, therefore S is also an ideal of S + T.

Example 11:

If U, V are ideals of a ring R, let

$$U + V = \{u + v : u \in U, v \in V\}.$$

Prove that U + V is also an ideal of R.

Solution:

Let $u_1 + v_1 \in U + V$ and $u_2 + v_2 \in U + V$. Then

$u_1 + u_2 \in U$ and $v_1 + v_2 \in V$.

We have $(u_1 + v_1) - (u_2 + v_2) = (u_1 - u_2) + (v_1 + v_2)$.

Since U is an ideal of R, therefore

$u_1, u_2 \in U \Rightarrow u_1 - u_2 \in U$.

Similarly V is also an ideal of R, therefore

$v_1, v_2 \in V \Rightarrow v_1 - v_2 \in V$.

$\therefore (u_1 - u_2) + (v_1 - v_2) \in U + V$.

$\therefore (u_1 + v_1) - (u_2 + v_2) \in U + V$.

$\therefore$ U + V is a subgroup of the additive group of R.

Now let r be any element of R and u + v be any element of U + V where $u \in U, v \in V$.

Then $r(u + v) = ru + rv$

$\in U + V$ since $r \in R$, $u \in U$ and U is an ideal

$\Rightarrow$ $ru \in U$ and similarly $rv \in V$

Similarly $(u + v)r = ur + vr \in U + V$ since $ur \in U$, $vr \in V$.

Hence U + V is an ideal of R.

Example 12:

Verify the following for being true or false.

1. *The set of all positive rationals is a subring of the ring of all rational numbers.*
2. *A subring of any field is a field.*
3. *Any subring of the ring of integers, Z, is an ideal of Z.*

Solution:

1. Let $(Q, +, \cdot)$ be the ring of all rational numbers and Q_+ be the set of all positive rational numbers.

Then Q_+ is not a subring of the ring (Q, +, ·). Obviously the rational number 0 which is the zero element of the ring of all rational numbers does not belong to Q_+. Therefore Q_+ is not a subring of the ring (Q, +, ·). Hence the given statement is false.

2. The statement that a subring of any field is a field is false. For example consider the field of rational numbers (Q, +, ·). The set of integers I is a subring of the field of rational numbers because $a \in I, b \in I \Rightarrow a - b \in I$ and $ab \in I$.

But the subring (I, + , ·) of the field (Q, + , ·) is not a field. The only element in I which possess multiplicative inverse are 1 and –1 while in a field every non-zero element must possess multiplicative inverse. Hence (I, + , ·) is not a field and so the given statement is false.

3. The statement that any subring of the ring of integers, Z, is an ideal of Z is true.

If S is the zero subring of Z, then obviously S is an ideal of Z.

If S is any subring other than the zero subring of the ring of integers Z and m is the smallest positive integer belonging to S, then we S = {xm : x is an integer}.

Now show that S is an ideal of the ring Z.

Hence any subring of the ring of integers, Z, is also an ideal of Z.

2

VECTOR SPACE

2.1 INTRODUCTION

We have already introduced Groups and Rings Theory. The third algebraic model which we are about to consider Vector Space can in large part, trace its origins to topics in Geometry and Physics.

Vector Space our their importance to the fact that so many models arising in the solution of specific problems turn out to be vector spaces for this reason the basic concept introduced in them have a certain university and are ones we encounter, and keep encountering, in so many diverse contexts. Among these fundamental notation are those of linear dependence, basis and dimension which will be developed in this chapter. These are potent and effective tools in all branches of mathematics. We shall make immediate and free use of these in many key places in next chapter which treats the theory of fields.

2.2 ELEMENTARY BASIC CONCEPT

Definition: A non-empty set V is said to be a vector space over a field F. It $\forall$ is an abelian group under an operation which we denoted by +, and if for every a $\in$ F, v $\in$ V there is defined an element, written av, in V subject to:

(i) α (V + w) = dv + dw

(ii) $(\alpha + \beta)v = \alpha v + \beta v$

(iii) $\alpha(\beta v) = (\alpha\beta)v$

(iv) 1.v = v.

for all $\alpha, \beta \in$ F, v w $\in$ V (where the 1 represents the unit element of F under multiplication).

Note: In axiom (i) above the + is that of V, whereas, in axiom (ii) on the left hand side it is that of F and on the right hand side it is that of V.

We shall consistly use the following notations:

(a) F will be a field.

(b) Lower Case Greek Letters will be elements of F; we shall often refer to element of F an scalars.

(c) Capital Latin Letters will denote vector space over F.

(d) Lower Case Latin Letters will denote elements of vector space we shall often call elements of a vector space vectors.

OR

Definition: *Let (F, + .) be a field. The elements of F will be called* **scalars.** *Let V be a non-empty set whose elements will be called* **vectors.** *Then V is a vector space over the field F, if*

1. *There is defined an internal composition in V called* **addition of vectors** *and denoted by '+'. Also for this composition V is an abelian group.*
2. *There is an external composition in V over F called* **scalar multiplication** *and denoted multiplicatively i.e., $a\alpha \in V$ for all $a \in F$ and for all $\alpha \in V$. In other words V is closed with respect to scalar multiplication.*
3. *The two compositions i.e., scalar multiplication and addition of vectors satisfy the following postulates:*

(i) $a(\alpha + \beta) = a\alpha + a\beta \ \forall a \in F$ *and* $\forall \alpha, \beta, \in V$.

(ii) $(a + b)\alpha = a\alpha + b\alpha \ \forall a, b, \in F$ *and* $\forall \alpha \in V$.

(iii) $(ab)\alpha = a(b\alpha) \ \forall a, b \in F$ *and* $\forall \alpha \in V$.

(iv) $1\alpha = \alpha \ \forall \in V$ *and 1 is the unity element of the field F.*

When V is a vector space over the field F, we shall say that V (F) is a vector space. If the field F is understood we can simply say that V is a vector space.

(i) $\alpha + \beta \in V$ for all $\alpha, \beta, \in V$.

(ii) $\alpha + \beta = \beta + \alpha$ for all $\alpha, \beta, \in V$

(iii) $\alpha + (\beta + \gamma) = (\alpha + \beta) + \gamma$ for all $\alpha, \beta, \gamma \in V$.

(iv) There exists an element $0 \in V$ such that

$$0 + \alpha = \alpha \ \forall \alpha \in V.$$

The element $0 \in V$ will be called the *zero vector.* It is the additive identity in V.

(v) To every vector $\alpha \in V$ there exists a vector $-\alpha \in V$ such that $-\alpha + \alpha = 0$. Thus each vector should possess additive inverse. The vector $-\alpha$ is called the negative of the vector α.

Example 1:

Let V be the set of all pairs (x, y) of real numbers, and let F be the field of real numbers. Define

$$(x, y) + (x_1, y_1) = (x + x_1, y + y_1)$$

$$c\,(x, y) = (cx, y).$$

Show that with these operations V is not a vector space over the field of real numbers.

Solution:

The operation of addition of vectors defined by

$$(x, y) + (x_1, y_1) = (x + x_1, y + y_1),$$

obviously (V, +) is an abelian group.

$(a + b)\,\alpha = a\alpha + b\alpha \;\forall\; a, b \in F$ and $\alpha \in V$ fails.

Again, let $\alpha = (x, y)$ and $a, b, \in F$. We have

$$(a + b)\,\alpha = (a + b)\,(x, y) = ((a + b)\,x, y)$$
$$= (ax + bx, y).$$

Also $a\alpha + b\alpha = a\,(x, y) + b\,(x, y) = (ax, y) + (bx, y)$

$$= (ax + bx, y + y) = (ax + bx, 2y).$$

Thus in general $(a + b)\,\alpha + a\alpha + b\alpha$

Hence V is not a vector space over the field.

Example 2:

Let R be the field of real numbers and let P_n be the set of all polynomials (of degree at most n) over the field R. Prove that P_n is a vector space over the field R.

Solution:

Here P_n is the set of all polynomials of degree at most n over the field R. The set P_n also includes the zero polynomial. Thus, we have

Thus $P_n = \{f(x): f(x) = a_0 + a_1x + a_2x^2 + ... + a_nx^n,$

Here $a_0, a_1, a_2, ..., a_n \in R\}$.

If $f(x) = a_0 + a_1x + a_2x^2 + ... + a_nx^n$

$g(x) = b_0 + b_1x + b_2x^2 + ... + b_nx^n$

be any two members of P_n, then

$$f(x) + g(x) = (a_0 + b_0) + (a_1 + b_1) x + ... + (a_n + b_n) x^n.$$

is also a member of P_n because it is also a polynomial of degree at most n over the field **R**.

Thus P_n is closed under for addition of polynomials.

Also we know that addition of polynomials is commutative as well as associative. The zero polynomial 0 is a member of P_n and is identity for addition of polynomials.

Also if $f(x) = a_0 + a_1x + ... + a_nx^n \in P_n$,

then $-f(x) = - a_0 - a_1x -...-a_nx^n$ Î P_n because it is also a polynomial of degree at most n over the field **R**.

We have $- f(x) + f(x)$ = the zero polynomial.

$\therefore - f(x)$ is the inverse of f(x) for addition of polynomials.

Hence P_n is an abelian group under for addition of composition.

Now if $f(x) = a_0 + a_1x + a_2x^2 +...+ a_nx^n$ is any member of P_n and $c \in$ **R**, we define scalar multiplication cf(x) by the relation

$$cf(x) = ca_0 + (ca_1) x + (ca_2) x^2 + ... + (ca_n) x^n.$$

Obviously $cf(x) \in P_n$ because it is also a polynomial of degree at most n over the field **R**. Thus P_n is closed for scalar multiplication.

Now if a, b $\in$ **R** and f(x), g(x) $\in P_n$, we have

$$a [f(x) + g(x)] = af(x) + ag(x),$$

$$(a + b) f(x) = af(x) + bg(x)$$

and (ab) f(x) = a [bf(x)] as can be easily shown.

Also 1 f(x) = f(x) " f(x) $\in P_n$.

Hence P_n is a vector space over the field **R**.

Example 3:

Prove that the set of all vectors in a plane over the field of real numbers is a vector space.

Solution:

Let us consider V be the set of all vectors in a plane and let R be the field of real numbers whose elements will be scalars.

Further, let $\alpha, \beta \in V$. If $\alpha = \overrightarrow{AB}$ and $\beta = \overrightarrow{BC}$, then we define $\alpha + \beta = \overrightarrow{AB} + \overrightarrow{BC} = \overrightarrow{AC}$. Since $\overrightarrow{AC}$ is also a vector, therefore $\alpha, \beta \in V \Rightarrow \alpha + \beta \in V$ and thus V is closed for addition of vectors. We know that addition of vectors on the set V is commutative as well as associative. The zero vector

$0 = \overrightarrow{AA}$ is identity for addition of vectors. If $\alpha = \overrightarrow{AB}$, then the vectors $-\alpha = \overrightarrow{BA}$ is the additive inverse of α because $-\alpha + \alpha = \overrightarrow{BA} + \overrightarrow{AB} = \overrightarrow{BB} = \vec{O}$

Hence (V, +) is an abelian group.

If $\alpha \in V$ and $m \in R$ *i.e.*, m is any scalar, then the scalar multiplication $m\alpha$ is defined as a vector whose direction is that of α or opposite to that α according as m is + ive or – ive and $|m\alpha| = |m| \cdot |\alpha|$.

Since $m \in \mathbf{R}, \alpha \in V \Rightarrow m\alpha \in V$ therefore V is closed for scalar multiplication 3.

Now if $a, b \in \mathbf{R}$ and $\alpha, \beta, \in V$

$a(\alpha + \beta) = a\alpha + a\beta$, $(a + b)\alpha = a\alpha + b\alpha$ and $(ab)\alpha = 0(b\alpha)$

Also if α is any vector and 1 is the multiplicative identity of the field **R**, 1α is in the direction of the vector α and

$$|1\alpha| = |1|.|\alpha| = 1.|\alpha| = |\alpha|.$$

Hence V is a vector space over the field R.

Example 4:

The vector of all real valued continuous (differentiable or integrable) functions defined in some interval [0, 1].

Solution:

Let us consider V denote the set of all real valued continuous functions of x defined in the interval [0, 1]. Then V is a vector space over the field R of real numbers with vector addition and scalar multiplication then we have

$$(f + g)(x) = f(x) + g(x) \ \forall \ f, g \in V$$

and $$(af)(x) = af(x) \ \forall \ a \in R, \forall \ f \subset V.$$

As in rings, we should first prove that V is an abelian group with respect to addition composition.

V is closed with respect to scalar multiplication since af is also a real valued continuous function in [0, 1]. Here, we observe that

1. If $a \in R$ and $f, g \in V$, then

$$[a(f + g)](x) = a[(f + g)(x)] = a[f(x) + g(x)] = af(x) + ag(x)$$
$$= (af)(x) + (ag)(x) = (af + ag)(x).$$

$\therefore a(f + g) = af + ag$.

2. If $a, b \in R$ and $f \in V$, then

$[(a + b) f] (x) = (a + b) f(x) = af(x) + bf(x) = (af) (x) + (bf) (x)$
$= (af + bf) (x).$

$\therefore (a + b) f = af + bf.$

3. If $a, b \in R$ and $f \in V$, then

$[(ab) f] (x) = (ab) f(x) = a [bf(x)] = a [(bf) (x)] = [a (bf)] (x):$

$\therefore (ab) f = a (bf).$

4. If 1 is the unity element of R and $f \in V$, then

$(1f) (x) = 1 f(x) = f(x).$

$\therefore 1f = f.$

Hence V is a vector space over R.

Example 5:

The set V of all m × n matrices with their elements as real numbers is a vector space over the field F of real numbers with respect to addition of matrices as addition of vectors and multiplication of a matrix by a scalar as scalar multiplication.

Solution:

We know that V is an abelian group w.r.t. addition of matrices and O the null matrix additive identity.

Again, if $a \in F$ and $\alpha \in V$ (*i.e.*, α is a matrix of the type $m \times n$ with elements as real numbers), then $a\alpha \in V$ because $a\alpha$ is also a matrix of the type $m \times n$ with elements as real numbers. Therefore V is closed with respect to scalar multiplication. Also from our study of matrices we observe that

(i) $a (\alpha + \beta) = a\alpha + a\beta \ \forall a \in F$ and $\forall \alpha, \beta \in V$.

(ii) $(a + b) \alpha = a\alpha + b\alpha \ \forall a, b, \in F$ and $\forall \alpha \in V$

(iii) $(ab) \alpha = a(b\alpha) \quad \forall a, b \in F$ and $\forall \alpha \in V$.

(iv) $1\alpha = \alpha \ \forall \alpha \in V$ where 1 is the unity element of the field F of real numbers.

Hence V (F) is a vector space.

Note: If V is the set of all m × n matrices with their elements as rational numbers and F is the field of real numbers, then V will not be closed with respect to scalar multiplication. For $\sqrt{3} \in F$ and if $\alpha \in V$, then $\sqrt{3}\,\alpha \notin V$ because the elements of the matrix $\sqrt{3}\,\alpha$ will not be rational numbers. Therefore V (F) will not be vector space.

Example 6:

Let $V = R^2 = \{(a_1, a_2): a_1, a_2 \in R\}$ and $F = R$.

Define the addition and scalar multiplication in R^2 as follows:

$$(a_1, a_2) + (b_1, b_2) = (a_1 + b_1, a_2 + b_2)$$

Show that $\boldsymbol{R^2}$ is a vector space over $\boldsymbol{R}$.

Solution:

Do yourself.

Example 7:

The vector space of all polynomials over a field F.

Solution:

Let us denote F(x) is the set of polynomial in an indeterminate x over a field F. Then F[x] is a vector space over the field F with respect to addition of two polynomials as addition of vectors and the product of a polynomial by a constant polynomial (*i.e.* by an element of F) as scalar multiplication.

Let $f(x) = \Sigma\, a_i x^i = a_0 + a_1 x + a_2 x^2 + a_3 x^3 + \ldots$

$g(x) = \Sigma\, b_i x^i = b_0 + b_1 x + b_2 x^2 + b_2 x^2 + \ldots$

and $h(x) = \Sigma\, c_i x^i = c_0 + c_1 x + c_2 x^2 + c_3 x^2 + \ldots$

be any arbitrary members of F[x].

Equality of Two Polynomials. We define f(x) = g(x) if and only if $a_1 = b_1$ for each $l = 0, 1, 2, \ldots$.

Addition Composition in F[x]. We define

$$f(x) + g(x) = (a_0 + b_0) + (a_1 + b_1)\, x + (a_2 + b_2)\, x^2 + \ldots$$

$$= \Sigma\, (a_1 + b_1)\, x^i.$$

Since $a_0 + b_0, a_1 + b_1, a_2 + b_2, \ldots$ are all elements of F, therefore f(x)+ g(x) $\in$ F[x] and thus F[x] is closed with respect to addition of polynomials.

Scalar Multiplication in F[x] Over F. If k is any scalar *i.e.*, k $\in$ F, we define

$$kf(x0 = ka_0 + (ka_1)\, x + (ka_2)\, x^2 + (ka_3)\, x^2 + \ldots$$

$$= \Sigma\, (ka_1)\, x^i.$$

Since $ka_0, ka_1, ka_2, \ldots$ are all elements of F, therefore kf (x) $\in$ F[x] and thus F[x] is closed with respect to scalar multiplication.

Now, we shall show that F[x] is a vector space for these two compositions.

Commutativity of Addition in F[x]. We have

$$f(x) + g(x) = (a_0 + b_0) + (a_1 + b_1) x + (a_2 + b_2) x^2 + ...$$
$$= (b_0 + a_0) + (b_1 + a_1) x + (b_2 + a_2) x^2 + ...$$

[$\because$ addition in the field F is commutative]

$$= g(x) + f(x).$$

Associativity of addition in F[x]. We have

$$[f(x) + g(x)] + h(x) = \Sigma (a_i + b_i) x^i + \Sigma c_i x^i$$
$$= \Sigma [(a_i + b_i)c_i x^i + \Sigma [a_i + (b_i + c_i)] x^i$$
$$= \Sigma a_i x^i + \Sigma (b_i + c_i) x^i = f(x) + [g(x) + h (x)].$$

Existence of additive identity in F[x]. Let 0 denote the zero polynomial over the field F *i.e.*,

$$0 = 0 + 0x + 0x^2 + 0x^2 + ...$$

Then $0 \in F[x]$ and $0 + f(x) = f(x)$.

$\therefore$ the zero polynomial 0 is the additive identity.

Existence of additive inverse of each member of F[x]. Let $-f(x)$ be the polynomial over the field F defined as

$$-(x) = -a_0 + (-a_1) x + (-a_2) x^2 + ...$$
$$= -a_0 - a_1x - a_2x^2 - ... .$$

Then $-f(x) \in F[x]$ and we get

$-f(x) + f(x) = 0$ *i.e.*, the zero polynomial.

$\therefore -f(x)$ is the additive inverse of f(x).

Thus F[x] is an abelian group with respect to addition of polynomials.

Now for the operation of scalar multiplication we make the following observations.

1. If $k \in F$, then

$$k [f(x) + g(x)] = k [(a_0 + b_0) + (a_1 + b_1) x + (a_2 + b_2) x^2 + ...]$$
$$= k (a_0 + b_0) + k (a_1 + b_1) x + k (a_2 + b_2) x^2 + ...$$
$$= (ka_0 + kb_0) + (ka_1 + kb_1) x + (ka_2 + kb_2) x^2 + ...$$
$$= [ka_0 + (ka_1) x + (ka_2) x^2 + ...] + [kb_0 + (kb_1) x + (kb_2) x^2 + ...]$$
$$= k (a_0 + a_1x + a_2x^2 + ...) + k (b_0 + b_1x + b_2x^2 + ...)$$
$$= kf (x) + kg (x).$$

2. If $k_1, k_2 \in F$, then

$(k_1 + k_2) f(x) = (k_1 + k_2) a_0 + [(k_1 + k_2) a_1] x$
$+ [(k_1 + k_2) a_2] x^2 + ...$

$= (k_1a_0 + k_2a_0) + (k_1a_1 + k_2a_1) x + (k_1a_2 + k_2a_2) x^2 +...$

$= [k_1a_0 + (k_1a_1) x + (k_1a_2) x^2 +...]$
$+ [k_2a_0 + (k_2a_1) x + (k_2a_2) x^2 + ...]$

$= k_1 (a_0 + a_1x + a_2x^2 + ...) + k_2 (a_0 + a_1x + a_2x^3 + ...)$

$= k_1 f(x) + k_2f(x).$

3. If $k_1, k_2 \in F$, then

$(k_1k_2) f(x) = (k_1k_2) a_0 + [(k_1k_2) a_1] x + [(k_1k_2) a_2] x^2 + ...$

$= k_1 (k_2a_0) + [k_1 (k_2a_1)] x + [k_1 (k_2a_2)] x^2+...$

$= k_1 [k_2a_0 + (k_2a_1) x + (k_2a_2 x^2 + ...]$

$= k_1 [k_2 f(x)].$

4. If 1 is the unity element of the field F, then

$1 f(x) = (1a_0) + (1a_1) x + (1a_2) x^2 + ...$

$= a_0 + a_1x + a_2x^2 + ... = f(x).$

Hence F[x] is a vector space over the field F.

Example 8:

A field K can be regarded as a vector space over any subfield F of K.

Solution:

We know that K is the set of vectors. Addition of vectors is the addition composition in the field K. Since K is a field, therefore (K, +) is an abelian group. Further the elements of the subfield F constitute the set of scalars. The composition of scalar multiplication is the multiplication composition in the field K. K is a field, therefore $a\alpha \in K \ \forall \ a \in F$ and $\forall \ \alpha \in K$ because both a and α are elements of K. If 1 is the unity element of K, then 1 is also the unity element of the subfield F.

(i) $a (\alpha + \beta) = a\alpha + a\beta \ \forall \ \alpha \in F$ and $\forall \ \alpha, \beta \in K$. (left distributive law).

(ii) $(a + b) \alpha = a\alpha + b\alpha \ \forall \ \alpha, \beta \in F$ and $\forall \ \alpha \in K$. (right distributive law).

(iii) $(ab) \alpha = a (b\alpha) \ \forall \ a, b \in F$ and $\forall \ \alpha \in K$. (associative law)

(iv) $1\alpha = \alpha \ \forall \ \alpha \in K$ and 1 is the unit element of the subfield F.

Since 1 is also the unity element of the field K, therefore $1\alpha = \alpha \ \forall \ \alpha \in K$. Hence K(F) is a vector space.

Example 9:

The vector of all ordered n-tuples over a field F.

Solution:

Let F be a field. An ordered set $a = (a_1, a_2, a_3, ..., a_n)$ of n element of F is called an *n-tuple* over F. Let V be the totality of all ordered *n-tuples* over F *i.e.*, let

$$V = \{(a_1, a_2, ..., a_n) : a_1, a_2, a_3, ..., a_n \in F\}.$$

Now, we shall give a vector space structure to V over the field F. Here this we define the following terms:

Equality of two n-tuples. Two elements $\alpha = (a_1, a_2, ..., a_n)$ and $\beta = (b_1, b_2, ..., b_n)$ of V are said to be equal if and only if

$a_i = b_i$ for each i = i, 2,..., n.

Addition composition in V. We define

$\alpha + \beta = (a_1 + b_1, a_2 + b_2, ..., a_n + b_n)$

$\forall\ \alpha = (a_1, a_2 ..., a_n), \beta = (b_1, b_2, ..., b_n) \in V.$

Since $a_1 + b_1, a_2 + b_2, ..., a_n + b_n$ are all elements of F, therefore $\alpha + \beta \in V$ and thus V is closed with respect to addition of n-tuples.

Scalar multiplication in V over F. We define

$a\alpha = (aa_1, aa_2, ..., aa_n)\ \forall\ \alpha \in F, \forall\ \alpha = (a_1, a_2, ..., a_n \in V.$

Since $aa_1, aa_2, ..., aa_n$ are all elements of F, therefore $a\alpha \in V$ and thus V is closed with respect to scalar multiplication.

Now we shall see that V is a vector space for these two compositions.

Associativity of addition in V. We have

$$(a_1, a_2, ..., a_n) + [(b_1, ..., b_n) + (c_1, c_2, ..., c_n)]$$
$$= (a_1, a_2, ..., a_n) + (b_1 + c_1, b_2 + c_2, ..., b_n + c_n)$$
$$= (a_1 + [b_1 + c_1], a_2 + [b_2 + c_2], ..., a_n + [b_n + c_n])$$
$$= ([a_1 + b_1] + c_1, [a_2 + b_2] + c_2, ..., [a_n + b_n] + c_n]$$
$$= (a_1 + b_1, a_2 + b_2, ..., a_n + b_n) + (c_1, c_2, ..., c_n)$$
$$= [(a_1, a_2, ..., a_n) + (b_1, b_2, ..., b_n)] + (c_1, c_2, ..., c_n).$$

Commutativity of addition in V. We have

$$(a_1, a_2, ..., a_n) + (b_1, b_2, ..., b_n) = (a_1 + b_1, a_2 + b_2, ..., a_n + b_n)$$
$$= (b_1 + a_1, b_2 + a_2, ..., b_n + a_n) = (b_1, b_2, ..., b_n) + (a_1, a_2, ..., a_n).$$

Existence of additive identity in V. We have

$(0, 0, \ldots, 0) \in V$. Also if $(a_1, a_2, \ldots, a_n) \in V$, then

$(a_1, a_2, \ldots, a_n) + (0, 0, \ldots, 0) = (a_1 + 0, a_2 + 0, \ldots, a_n + 0)$

$= (a_1, a_2, \ldots, a_n)$.

$\therefore (0, 0, \ldots, 0)$ is the additive identity in V.

$(a_1, a_2, \ldots, a_n) \in V$, then $(-a_1, -a_2, \ldots, -a_n) \in V$.

Also we have $(-a_1, -a_2, \ldots, -a_n) + (a_1, a_2, \ldots, a_n)$

$= (-a_1 + a_1, -a_2 + a_2, \ldots, -a_n + a_n) = (0, 0, \ldots, 0)$.

$\therefore (-a_1, -a_2, \ldots, -a_n)$ is the additive inverse of $(a_1, a_2, \ldots, a_n)$.

Thus V is an abelian group with respect to addition. Further we observe that

1. If $a \in F$ and $\alpha (a_1, a_2, \ldots, a_n)$, $\beta = (b_1, b_2, \ldots, b_n) \in V$, then $a (\alpha + \beta)$
$= a (a_1 + b_1, a_2 + b_2, \ldots, a_n + b_n)$
$= (a [a_1 + b_1], a [a_2 + b_2], \ldots, a [a_n + b_n])$
$= (aa_1, ab_1, aa_2 + ab_2, \ldots, aa_n + ab_n)$
$= (aa_1, aa_2, \ldots, aa_n) + (ab_1, ab_2, \ldots, b_n)$
$= a(a_1, a_2, \ldots, a_n) + a(b_1, b_2, \ldots, b_n) = a\alpha + a\beta$.
2. If $a, b \in F$ and $a = (a_1, a_2, \ldots, a_n) \in V$, then
$(a + b) \alpha = ([a + b] a_1, [a + b] a_2, \ldots, [a + b] a_n)$
$= (aa_1 + ba_1, aa_2 + ba_2, \ldots, aa_n + ba_n)$
$= (aa_1, aa_2, \ldots, aa_n) + (ba_1, ba_2, \ldots, ba_n)$
$= a (a_1, a_2, \ldots, a_n) + b (a_1, a_2, \ldots, a_n) = a\alpha + b\alpha$.
3. If $a, b, \in F$ and $\alpha - (a_1, a_2, \ldots, a_n) \subset V$, then
$(ab) \alpha - ([ab] a_1, [ab] a_2, \ldots, [ab] a_n) - (a [ba_1], a [ba_2], \ldots, a [ba_n])$
$= a (ba_1, ba_2, \ldots, ba_n) = a [b (a_1, a_2, \ldots, a_n)] = a (b\alpha)$.
4. If 1 is unity element of F and $\alpha = (a_1, a_2, \ldots, a_n) \in V$, then $1\alpha = (1a_1, 1a_2, \ldots, 1a_n) = (a_1, a_2, \ldots, a_n) - \alpha$.

Hence V is a vector space over F. *The vector space of all ordered n-tuples over F will be denoted by $V_n(F)$. Sometimes we also denote it by F^n.* Here the zero vector *i.e.*, 0 is the n-tuple $(0, 0, \ldots, 0)$.

Example 10:

Show that a field F may be considered as a vector space over F if scalar multiplication is identified with field multiplication.

Solution:

Let us consider (F, +, ·) be a field. Take F as the set of vectors and also as the set of scalars. Take the addition operation on the field F as the addition of vectors and the multiplication operation on the field F as the operation of scalar multiplication *i.e.*, multiplication of a vector by a scalar. Then F is a vector space over F as shown as follows:

Since (F, +, ·) is a field, therefore (F, +) is an abelian group.

Again, if a, b are any scalars *i.e.*, a, b, $\in$ F and α, β, are any vectors *i.e.*, α, β, $\in$ F then we have

$$a(\alpha + \beta) = a\alpha + a\beta \text{ and } (a + b)\alpha = a\alpha + b\alpha.$$

(by distribution law)

Also (ab) α = a (bα) because the multiplication on the field F is associative.

Also if 1 is the unity element of the field F and α is any vector *i.e.*, $\alpha \in$ F, then 1α = α.

Hence F (F) is a vector space.

Example 11:

Show that the complex field C is a vector space over the real field R.

Solution:

Let us consider C as the set of vectors and R as the set of scalars.

Take the addition of complex numbers as the addition of vectors. Then (C, +) is an abelian group.

Again, let a be any scalar *i.e.*, a $\in$ R and α be any vector *i.e.*, $\alpha \in$ C. Now a $\in$ R $\Rightarrow$ a $\in$ C. Thus, both a and α are in C. Regard the composition of scalar multiplication as the multiplication of a and α in the field. Since C is a field, therefore a$\alpha \in$ C and thus C is closed for scalar multiplication.

The real number 1 which is the unity element of the field R is also the unity element of the field C.

Now let α, β be any vectors *i.e.*, α, $\beta \in$ C and a, b be any scalars *i.e.* a, b $\in$ R. Since R $\subset$ C, therefore, we have

$$a, b \in R \Rightarrow a, b \in C$$

We have a($\alpha + \beta$) = aα + aβ and a(a + b) α = aα + bα. (distributive laws in the field C).

Also (ab) α = a (bα) because the multiplication on the field C is associative.

Also $1\alpha = \alpha$ because 1 is also the multiplicative identity in the field C.

Hence C is a vector space over R.

Example 12:

How many elements are there in the vector space of polynomials of degree at most in which the coefficients are the elements of the field I (p) over the field I (p), p being a prime number?

Solution:

Here we know that the field I (p) is the field

$(\{0, 1, 2,..., p-1)\}, +_{p,} \times_{p})$.

The number of distinct elements in the field I (p) is p.

If f(x) is a polynomial of degree at most n over the field I (p), then we have

$f(x) = a_0 + a_1x + a_2x^2 + ... + a_nx^n$, where $a_0, a_1, a_2,..., a_n \in$ I (p).

Now in the polynomial f(x), the coefficient of each of the n + 1 terms $a_0, a_1x, a_2x^2,..., a_nx^n$ can be filled in p ways because any of the p elements of the field I (p) can be filled there.

Thus, we can have p × p × p × ... upto (n + 1) times *i.e.*, p^{n+1} distinct polynomials of degree at most n over the field I (p). Hence if P_n is the vector space of polynomials of degree at most n in which the coefficients are the elements of the field I (p) over the field I (p), then P_n has p^{n+1} distinct elements.

2.3 GENERAL PROPERTIES OF VECTOR SPACES

Theorem 1:

Let V(F) be a vector space and 0 be the zero vector of V. Then

(i) $a0 = 0 \ \forall \ \alpha \in F$.

(ii) $0\alpha = 0 \ \forall \ \alpha \in V$.

(iii) $a(-\alpha) = -(a\alpha) \ \forall \ \alpha \in F, \forall \ \alpha \in V$.

(iv) $(-a)\, a = -(a\alpha) \ \forall \ \alpha \in F, \forall \ \alpha \in V$.

(v) $a(\alpha - \beta) = a\alpha - a\beta \ \forall \ \alpha \in F$ and $\forall \ \alpha, \beta \in V$.

(vi) $a\alpha = 0 \Rightarrow a = 0$ or $\alpha = 0$.

Proof:

The proofs is very easy and follows the lines of the analogues result proved for rings for this reason we give it briefly and with few explanations. Here

(i) We have $a0 = a(0 + 0)$ $[\because 0 = 0 + 0]$

$= a0 + a0.$

$\therefore 0 + a0 + a0$ $[\because a0 \in V \text{ and } 0 + a0 = a0]$

Now V is an abelian group with respect to addition.

Hence by right cancellation law in V, we get $0 = a0$.

(ii) We have $0\alpha = (0 + 0)\alpha$ $[\because 0 = 0 + 0]$

$= 0\alpha + 0\alpha.$

$\therefore 0 + 0\alpha = 0\alpha + 0\alpha$. $[\because 0\alpha \in V \text{ and } 0 + 0\alpha = 0\alpha]$

Now V is an abelian group with respect to addition of vectors. Therefore by right cancellation law in V, we get $0 = 0\alpha$.

(iii) We have $a[\alpha + (-\alpha)] = a\alpha + a(-\alpha)$

$\Rightarrow a0 = a\alpha + a(-\alpha)$

$\Rightarrow 0 = a\alpha + a(-\alpha)$ $[\because a0 = 0]$

$\Rightarrow$ ($-a\,\alpha$ is the additive inverse of $a\alpha$)

$\Rightarrow a(-\alpha) = -(a\alpha).$

(iv) We have $[a + (-a)]\alpha = a\alpha + (-a)\alpha$

$\Rightarrow 0\alpha = a\alpha + (-a)\alpha$

$\Rightarrow 0 = a\alpha + (-a)\alpha$

$\Rightarrow (-a)\alpha$ is the additive inverse of $a\alpha$

$\Rightarrow (-a)\alpha = -(a\alpha).$

(v) We have $(\alpha - \beta) = a[\alpha + (-\beta)] = a\alpha + a(-\beta)$

$= a\alpha + [-(a\beta)$ $[\because a(-\beta) = -(a\beta)]$

$= a\alpha - a\beta.$

(vi) Let $a\alpha = 0$ and $a \neq 0$. Then a^{-1} exists because a is a non-zero element of the field F.

$\therefore a\alpha = 0 \Rightarrow a^{-1}(a\alpha) = a^{-1}0 \Rightarrow (a^{-1}a)\alpha = 0 \Rightarrow 1\alpha = 0 \Rightarrow \alpha = 0.$

Again let $a\alpha = 0$ and $\alpha \neq 0$. Then to prove that $a = 0$. Suppose $\alpha \neq 0$. Then a^{-1} exists.

$\therefore a\alpha = 0 \Rightarrow a^{-1}(a\alpha) = a^{-1}0 \Rightarrow (a^{-1}a)\alpha = 0 \Rightarrow 1\alpha = 0 \Rightarrow \alpha = 0.$

Thus we get a contradiction that α must be a zero vector. Therefore a must be equal to 0. Hence $\alpha \neq 0$ and $a\alpha = 0 \Rightarrow a = 0$.

Theorem 2:

Let V (F) be a vector space. Then

(i) *If* $a, b \in F$ *and* α *is anon-zero element of* V, *we have*

$$a\alpha = b\alpha \Rightarrow a = \beta.$$

(ii) *If* $\alpha, \beta \in V$ *and* α *is a non-zero element of* F, *we have*

$$a\alpha = a\beta \Rightarrow a = b.$$

Proof:

(i) We have $a\alpha = b\alpha \Rightarrow a\alpha - b\alpha = 0 \Rightarrow (a - b)\,\alpha = 0$.
But $\alpha \neq 0$. Therefore $(a - b)\,\alpha = 0 \Rightarrow a - b = 0 \Rightarrow a - b$.

(ii) We have $a\beta = a\beta \Rightarrow a\alpha - a\beta = 0 \Rightarrow a(\alpha - \beta) = 0$.
But $a \neq 0$. Therefore $a(\alpha - \beta) = 0 \Rightarrow \alpha - \beta = 0 \Rightarrow \alpha = \beta$.

2.4 VECTOR SUBSPACES

Definition: *Let V a vector space over the field F and* $W \subseteq V$. *Then W is called a subspace of V if W itself is a vector space over F with respect to the operations of vector addition and scalar multiplication in V.*

Theorem 1:

The necessary and sufficient condition for a non empty subset W of a vector space V (F) to be a subspace of V is

$$a, b \in F \text{ and } \alpha, \beta \in W \Rightarrow a\alpha + b\beta \in W.$$

Proof:

The condition necessary. If W is a subspace of V, then W must be closed under scalar multiplication and vector addition.

Therefore $\quad a \in F, \alpha \in W \Rightarrow a\alpha \in W$

and $\quad b \in F, \beta \in W \Rightarrow b\beta \in W$.

Now $a\alpha \in W, b\beta \in W \Rightarrow a\alpha + b\beta \in W$. Hence the condition is necessary.

The condition is sufficient. Now suppose W is a non-empty subset of V satisfying the given condition *i.e.*, $a, b, \in F$ and

$$\alpha, \beta \in W \Rightarrow a\alpha + b\beta \in W.$$

Taking $a = 1, b = 1$, we see that if $\alpha, \beta \in W$ then

$1\alpha + 1\beta \in W$

$\Rightarrow \alpha + \beta \in W$. $\qquad [\because \alpha \in W \Rightarrow \alpha \in V \text{ and } 1\alpha = \alpha \text{ in } V]$

Thus W is closed under vector addition.

Now taking $a = -1, b = 0$, we see that if $\alpha \in W$, then

$(-1)\,\alpha + a\alpha \in W$ [in place of β we have taken α]

$\Rightarrow -(1\alpha) + 0 \in W \Rightarrow -\alpha \in W.$

Thus the additive inverse of each element of W is also in W.

Taking a = 0, b = 0, we see that if a ∈ W, then

$0\alpha + 0\alpha \in W \Rightarrow 0 + 0 \in W \Rightarrow 0 \in W.$

Thus the zero vector of V belongs to W. It will also be the zero vector of W.

Since the elements of W are also the elements of V, therefore vector addition will be associative as well as commutative in W. Thus W is an abelian group with respect to vector addition.

Now taking β = 0, we see that if a, b, ∈ F and α ∈ W, then

$$a\alpha + b0 \in W \text{ i.e. } a\alpha + 0 \in W \text{ i.e., } a\alpha \in W.$$

Thus W is closed under scalar multiplication.

The remaining postulates of a vector space will hold in W since they hold in V of which W is a subset. Hence W (F) is a subspace of V (F).

Note: If we are to prove that a subset W of a vector space V is a subspace of V, then it is sufficient to prove that

$a, b \in F$ and $\alpha, \beta, \in W \Rightarrow a\alpha + b\beta \in W.$

Theorem 2:

The necessary and sufficient conditions for a non-empty subset W of a vector space V (F) to be a subspace of V are

(i) $\alpha \in W, \beta \in W \Rightarrow \alpha - \beta \in W.$

(ii) $a \in F, \alpha \in W \Rightarrow a\alpha \in W.$

Proof:

The conditions are necessary. Let if W is a subspace of V, then W is an abelian group with respect to vector addition. Therefore α ∈ W, β ∈ W ⇒ α – β ∈ W. Also W must be closed under scalar multiplication. Therefore the condition (ii) is also necessary.

The conditions are sufficient. Again let W is a non-empty subset of V satisfying the two given conditions. From condition (i), we have

$$\alpha \in W, \alpha \in W \Rightarrow \alpha - \alpha \in W \Rightarrow 0 \in W.$$

Hence, the zero vector of V belongs to W and it will also be the zero vector of W.

Now $0 \in W, \alpha \in W \Rightarrow 0 - \alpha \in W \Rightarrow -\alpha \in W.$

Thus the additive inverse of each element of W is also in W.

Again $\alpha \in W, \beta \in W \Rightarrow \alpha \in W, -\beta \in W$

$\Rightarrow \quad \alpha - (-\beta) \in W \Rightarrow \alpha + \beta \in W.$

Thus, W is closed with respect to vector addition.

Since the elements of W are also the elements of V, therefore vector addition will be commutative as well as associative in W. Hence W is an abelian group under vector addition. Also from condition (ii), W is closed under scalar multiplication. The remaining postulates of a vector space will hold in W since they hold in V which is a superset of W. Hence W is a subspace of V.

Theorem 3:

The necessary and sufficient condition for a non-empty subset W of a vector space V (F) to be a subspace of V is that W is closed under addition and scalar multiplication in V.

Proof:

If W itself is a vector space over F with respect to vector addition and scalar multiplication in V, then W must be closed with respect to these two compositions.

The condition is sufficient. Now let that W is a non-empty subset of V and W is closed under vector addition and scalar multiplication in V.

Let $\alpha \in W$, if 1 is the unity element of F, then $-1 \in F$. Now W is closed under scalar multiplication. Therefore, we have

$$-1 \in F, \alpha \in W \Rightarrow (-1)\alpha \in W \Rightarrow -(1\alpha) \in W$$

$\Rightarrow \quad -\alpha \in W \qquad [\because \alpha \in W \Rightarrow \alpha \in V \text{ and } 1\alpha = \alpha \text{ in } V].$

Thus the additive inverse of each element of W is also in W.

Hence W is closed under vector addition.

Therefore $\alpha \in W, -\alpha \in W \Rightarrow \alpha + (-\alpha) \in W$

$\Rightarrow 0 \in W$ where 0 is the zero vector of V.

Hence the zero vector of V is also the zero vector of W. Since the elements of W are also the elements of V, hence, vector W is an abelian group with respect to vector addition. Also, we know that W is closed under scalar multiplication. The remaining postulates of a vector space will hold in W since they hold in V of which W is a subset.

Hence W itself is a vector space for the two compositions.

$\therefore$ W is a subspace of V.

Example 1:

The set W of ordered triads $(a_1, a_2, 0)$ where $a_1, a_2 \in F$ is a subspace of V_3 (F).

Solution:

Let us consider $\alpha = (a_1, a_2, 0)$ and $\beta = (b_1, b_2, 0)$ be any two elements of W. Then $a_1, a_2, b_1, b_2 \in F$. If a, b be any two elements of F, we have

$a\alpha + b\beta = a(a_1, a_2, 0) + b(b_1, b_2, 0) = (aa_1, aa_2, 0) + (bb_1, bb_2, 0)$

$= (aa_1 + bb_1, aa_2 + bb_2, 0) \in W$ since $aa_1 + bb_1, aa_2 + bb_2 \in F$ and the last co-ordinate of this triad is zero.

Hence W is a subspace of V_3 (F).

Example 2:

Let V be the vector space of all polynomials in an indeterminate x over a field F. Let W be a subset of V consisting of all polynomials of degree $\leq$ n. Then W is a subspace of V.

Solution:

Let α and β be any two elements of W. Then α, β, are polynomials over F of degree $\leq$ n. If a, b are any two elements of F, then $a\alpha + b\beta$ will also be a polynomial of degree $\leq$ n. Therefore $a\alpha + b\beta \in W$. Hence W is a subspace of V.

Example 3:

*Let **R** be the field of real numbers. Which of the following are subspaces of V_3 **(R)** ?*

(i) $\{(x, 2y, 3z): x, y, z \in \mathbf{R}\}$

(ii) $\{(x, x, x): x \in \mathbf{R}\}$.

(iii) $[(x, y, z): x, y, z$ are rational numbers$\}$?

Solution:

(i) Let $W = \{(x; 2y, 3z): x, y, z \in \mathbf{R}\}$.

Let $\alpha = (x_1, 2y_1, 3z_1)$ and $\beta = (x_2, 2y_2, 3z_2)$ be any two elements of W. Then $x_1, y_1, z_1, x_2, y_2, z_2$ are real numbers. If a, b are any two real numbers, then

$$a\alpha + b\beta = a(x_1, 2y_1, 3a_1) + b(x_2, 2y_2, 3z_2)$$

$= (ax_1 + bx_2, 2ay_1 + 2by_2, 3az_1 + 3bz_2)$

$= (ax_1 + bx_2, 2 [ay_1 + by_2], 3 [az_1 + bz_2])$

$\in$ W since $ax_1 + bx_2$, $ay_1 + by_2$, $az_1 + bz_2$ are real numbers.

Thus a, b $\in$ **R** and $\alpha, \beta \in W \Rightarrow a\alpha + b\beta \in W$.

$\therefore$ W is a subspace of V_3 (**R**).

(ii) Let W = {(x, x, x): x $\in$ **R**}.

Let $\alpha = (x_1, x_1, x_1)$ and $\beta = (x_2, x_2, x_2)$ be any two elements of W. Then x_1, x_2, are real numbers. If a, b, are any real numbers, then

$a\alpha + b\beta = a (x_1, x_1, x_1) + b (x_2, x_2, x_2)$

$= (ax_1 + bx_2, ax_1 + bx_2, ax_1 + bx_2) \in W$

since $ax_1 + bx_2 \in$ **R**.

(iii) Let W = {(x, y, z): x, y, z are rational numbers}.

Now $\alpha = (3, 4, 5)$ is an element of W. Also $a = \sqrt{3}$ is an element of R. But $a\alpha = \sqrt{3}(3, 4, 5) = (3\sqrt{3}, 4\sqrt{3}, 5\sqrt{3}) \notin W$ since $3\sqrt{3}$, $4\sqrt{3}$, $5\sqrt{3}$ are not rational numbers.

Therefore W is not closed under scalar multiplication. Hence W is not a subspace of V_2 (**R**).

Example 4:

If a_1, a_2, a_3 are fixed elements of a field F, then the set W of all ordered triads (x_1, x_2, x_3) of elements of F, such that $a_1x_1 + a_2x_2 + a_2x_2 = 0$. is a subspace of V_2 (F).

Solution:

Let $\alpha = (x_1, x_2, x_3)$ and $\beta = (y_1, y_2, y_3)$ be any two elements of W. Then $x_1, x_2, x_3, y_1, y_2, y_3$ are elements of F and are such that

$$a_1x_1 + a_2x_2 + a_3x_3 = 0 \quad ...(1)$$

$$a_1y_1 + a_2y_2 + a_3y_3 = 0 \quad ...(2)$$

If a, b, be any two elements of F, we have

$a\alpha + b\beta = a (x_1, x_2, x_3) + b (y_1, y_2, y_3)$

$= (ax_1, ax_2, ax_3) + (by_1, by_2, by_3) = (ax_1 + by_1, ax_2 + by_2, ax_3 + by_3)$.

Now $a_1 (ax_1 + by_1) + a_2 (ax_2 + by_2) + a_3 (ax_3 + by_3)$

$= a (a_1x_1 + a_2x_2 + a_3x_3) + b (a_1y_1 + a_2y_2 + a_3y_3)$

$= a0 + b0$ [by (1) and (2)]

$= 0$.

$\therefore a\alpha + b\beta = (ax_1 + by_1, ax_2 + by_2, ax_3 + by_3) \in W$.

Hence W is a subspace of V_3 (F).

2.5 ALGEBRA OF SUBSPACE

Theorem 1:

The intersection of any two subspaces W_1 and W_2 of a vector space V(F) is also a subspace of V(F).

Proof:

Since $0 \in W_1$ and W_2 both therefore $W_1 \cap W_2$ is not empty.

Let $\alpha, \beta \in W_1 \cap W_2$ and $a, b \in F$.

Now $\alpha \in W_1 \cap W_2 \Rightarrow \alpha \in W_1$ and $\alpha \in W_2$.

and $\beta \in W_1 \cap W_2 \Rightarrow \beta \in W_1$ and $\beta \in W_2$.

Since W_1 is a subspace, therefore

$a, b \in F$ and $\alpha, \beta, \in W_1 \Rightarrow a\alpha + b\beta \in W_1$.

Similarly $a, b, \in F$ and $\alpha, \beta, \in W_2 \Rightarrow a\alpha + b\beta \in W_2$.

Now $a\alpha + b\beta \in W_1$, $a\alpha + b\beta \in W_2 \Rightarrow a\alpha + b\beta \in W_1 \cap W_2$.

Thus $a, b, \in F$ and $\alpha, \beta, \in W_1 \cap W_2 \Rightarrow a\alpha + b\beta \in W_1 \cap W_2$.

Hence $W_1 \cap W_2$ is a subspace of V(F).

Note: The union of two subspaces of V(F) may not be a subspace of V(F). For example if **R** be the field of real numbers, then $W_1 = \{(0, 0, z); z \in \mathbf{R}\}$ are two subspaces of V2(R). We have $(0, 9, 3) \in W_1$ and $(0, 5, 0) \in W_2$.

But $(0, 0, 3) + (0, 5, 0) = (0, 5, 3) \notin W_1 \cup W_2$ since neither (0, 5, 3) $\in W_2$. Thus $W_1 \cup W_2$ is not closed under vector addition. Hence $W_1 \cup W_2$ is not a subspace of $V_2(\mathbf{R})$.

Theorem 2:

Arbitrary intersection of subspaces i.e., the intersection of any family of subspaces of vector space is a subspace.

Proof:

Suppose V (F) be a vector space and let $\{W_t: t \in T\}$ be any family of subspaces of V. And T is an index set such that $\forall\, t \in T$, Wt is a subspace of V.

Let $$U = \bigcap_{t \in T} W_t = \{x \in V: x \in W_t\ \forall\, t \in T\}$$

be the intersection of this family of subspaces of V. Then to prove that U is also a subspace of V.

We know that $U \neq \varnothing$, since at least the zero vector 0 of V is in $W_t \ \forall \ t \in T$.

Now let a, b, $\in$ F and α, β, be any two elements of $\bigcap_{t \in T} W_t$.

Then $\alpha, \beta \in W_t \ \forall \ t \in T$. Since each W_t is a subspace of V, therefore $a\alpha + b\beta \in W_i \ \forall \ t \in T$. Thus $a\alpha + b\beta \in \bigcap_{t \in T} W_t$.

Thus a, b, $\in$ F and $\alpha, \beta, \in \bigcap_{t \in T} W_t \Rightarrow a\alpha + b\beta \in \bigcap_{t \in T} W_t$.

Hence $\bigcap_{t \in T} W_t$ is a subspace of V(F).

Theorem 3:

The union of two subspace is a subspace if and only if one is contained in the other.

Proof:

Suppose W_1 and W_2 are two subspaces of a vector space V.

Let $W_1 \subseteq W_2$ of $W_2 \subseteq W_1$. Then $W_1 \cup W_2 = W_2$ or W_1. But W_1, W_2 are subspaces and therefore, $W_1 \cup W_2$ is also a subspace.

Conversely, suppose $W_1 \cup W_2$ is subspace.

To prove that $W_1 \subseteq W_1$ or $W_2 \subseteq W_1$.

Let us assume that W_1 is not a subset of W_2 and W_2 is also not a subset of W1.

Now W_1 is not a subset of $W_2 \Rightarrow \exists \ \alpha \in W_1$ and $\alpha \notin W_2$ ...(i)

and W_2 is not a subset of $W_1 \Rightarrow \exists \ \beta \in W_2$ and $\beta \notin W_2$ ...(ii)

From (i) and (ii), we have

$\alpha \in W_1 \cup W_2$ and $\beta \in W_1 \cup W_2$.

Since $W_1 \cup W_2$ is an subspace, therefore

$\alpha + \beta$ is also in $W_1 \cup W_2$.

But $\alpha + \beta \in W_1 \cup W_2 \Rightarrow \alpha + \beta \in W_1$ or W_2.

Suppose $\alpha + \beta \in W_1$. Since $\alpha \in W_1$ and W_2 is a subspace, therefore $(\alpha + \beta) - \alpha = \beta$ is in W_1.

But from (2), we have $\beta \notin W_1$. Thus we get a contradiction. Again suppose that $\alpha + \beta \in W_2$. Since $\beta \in W_2$ and W_2 is subspace, therefore

$(\alpha + \beta) - \beta = \alpha$ is in W_2. But from (1), we have a Ï W_2. Thus here also we get a contradiction. Hence either $W_1 \subseteq W_2$ or $W_2 \subseteq W_1$.

Smallest subspace containing any subset of V(F). Let V(F) be a vector space and S be any subset of V. If U is subspace of V containing S and is itself contained in every subspace of V containing S, then U is called the *smallest subspace* of V containing S. The smallest subspace of V containing S is also called the subspace of V generated or spanned by S and we shall denote it by the symbol {S} or by (S). It can be easily seen that the intersection of all the subspaces of V(F) containing S is the subspace of V(F) generated by S If {S) = V, then we say that V is spanned by S.

Example 1:

*Prove that the set of all solutions (a, b, c) of the equation $a + b + 2c = 0$ is a subspace of the vector space V_3 (**R**).*

Solution:

Let $W = \{(a, b, c): a, b, c \in \mathbf{R} \text{ and } a + b + 2c = 0\}$.

To prove that W is a subspace of V_3 (**R**) or $\mathbf{R}^3$.

Let $\alpha = (a_1, b_1, c_1)$ and $\beta = (a_2, b_2, c_2)$ be any two elements of W. Then

$$a_1 + b_1 + 2c_1 = 0 \quad \text{...(1)}$$

and $$a_2 + b_2 + 2c_2 = 0 \quad \text{...(2)}$$

If a, b be any two elements of **R**, we have

$$a\alpha + b\beta = a\,(a_1, b_1, c_1) + b\,(a_2, b_2, c_2)$$
$$= (aa_1, ab_1, ac_1) + (ba_2, bb_2, bc_2)$$
$$= (aa_1 + ba_2, ab_1 + bb_2, ac_1 + bc_2).$$

Now $(aa_1 + ba_2) + (ab_1 + bb_2) + 2\,(ac_1 + bc_2)$

$$= a\,(a_1 + b_1 + 2c_1) + b\,(a_2 + b_2 + 2c_2)$$
$$= a.0 + b.0 \quad \text{[from (1) and (2)]}$$
$$= 0.$$

$\therefore\ a\alpha + b\beta\ (aa_1 + ba_2, ab_1 + bb_2, ac_1 + bc_2) \in W$

Thus $\Rightarrow \alpha, \beta \in W$ and $a, b \in R \Rightarrow a\alpha + b\beta \in W$.

Hence W is a subspace of V_3(**R**).

Example 2:

Which of the following sets of vectors $\alpha = (a_1, a_2, \ldots, a_n)$ in R^n are subspaces of R^n $(n \geq 3)$?

(i) all α such that $a_1 \leq 0$;

(ii) all α such that a_3 is an integer;

(iii) all α such that $a_2 + 4a_2 = 0$.

Solution:

(i) Let $W = \{\alpha: \alpha \in \mathbf{R}^n \text{ and } a_1 \leq 0\}$.

If we take $a_1 = -3$, then $a_1 < 0$ and so

$$\alpha = (-3, a_2, \ldots, a_n) \in W.$$

Now if we take $a = -2$, then

$$a\alpha = (6, -2a_2, \ldots, -2a_2).$$

Since the first coordinate of $a\alpha$ is 6 which is > 0, therefore $a\alpha \notin W$.

Thus, $\alpha \in W$, $a \in R$ but $a\alpha \notin W$. Therefore W is not closed for scalar multiplication and so W is not a subspace of $\mathbf{R}^n$.

(ii) Let $W = \{\alpha: \alpha \in \mathbf{R}^n \text{ and } a_2 \text{ is an integer}\}$.

If we take $a_3 = 5$, then a_3 is an integer and so

$$a = (a_1, a_2, 5, \ldots, a_n) \in W.$$

Now if we take $a = \frac{1}{2}$, then

$$a\alpha = \left(\frac{1}{2}a_1, \frac{1}{2}a_2, 5/2, \ldots, \frac{1}{2}a_n\right).$$

Since the third coordinate of $a\alpha$ is 5/2 which is not in integer, therefore $a\alpha \notin W$.

Thus, $\alpha \in W$, $\alpha \in \mathbf{R}$ but $a\alpha \in W$. Therefore is not closed for scalar multiplication and so W is not a subspace of $\mathbf{R}^n$.

(iii) Let $\alpha = (a_1, \ldots, a_n)$ and $\beta - (b_1, \ldots, b_n)$ be any two members of W. Then $a_2 + 4a_2 = 0$ and $b_2 + 4b_3 = 0$.

If $a, b, \in \mathbf{R}$, then $a\alpha + b\beta = (aa_1 + bb_1, \ldots, aa_n + bb_n)$.

We have $(aa_2 + bb_2) + 4\,(aa_3 + bb_3)$

$$= a\,(a_2 + 4a_3) + b\,(b_2 + 4b_3) + b\,(b_2 + 4b_3 = a, 0 + b.0 = 0.$$

Thus according to the definition of W, $a\alpha + b\beta \in W$.

In this way $\alpha, \beta, \in W$ and $a, b, \in \mathbf{R} \Rightarrow a\alpha + b\beta \in W$. Hence W is a subspace of $\mathbf{R}^n$.

Example 3:

*Show that the set W of the elements of the vector space V_3 **(R)** of the form $(x + 2y, y, -x + 3y)$, where $x, y \in R$ is a subspace of V_3 **(R)**.*

Solution:

Let $W = \{(x + 2y, y, -x + 3y): x, y \in \mathbf{R}\}$.

To prove that W is a subspace of V_3 (**R**).

Let $\alpha = (x_1 + 2y, y_1, -x_1 + 3y_1)$ and $\beta = (x_2 + 2y_2, -x_2 + 3y_2)$ be any two elements of W.

If a, b be any two elements of **R**, we have

$a\alpha + b\beta = a(x_1 + 2y_1, y_1, -x_1 + 3y_1) + b(x_2 + 2y_2, -x_2 + 3y_2)$

$= (ax_1 + 2ay_1, ay_1, -ax_1 + 3ay_1) + (bx_2 + 2by_2, by_2, -bx_2 + 3by_2)$

$= (ax_1 + 2ay_1 + bx_2 + 2by_2, ay_1 + by_2, -ax_1 + 3ay_1 - bx_2 + 3by_2)$

$= ([ax_1 + bx_2] + 2[ay_1 + by_2], ay_1 + by_2, -[ax_1 + bx_2] + 3[ay_1 + by_2]$

which is in W because it is of the form $(x + 2y, y, -x + 3y)$.

Here in place of y we have $ay_1 + by_2$ and in place of x we have

$ax_1 + bx_2$.

Thus $\alpha, \beta \in W$ and $a, b \in \mathbf{R} \Rightarrow a\alpha + b\beta \in W$.

Hence W is a subspace of V_2 (**R**).

2.6 LINEAR COMBINATION OF VECTORS: LINEAR SPAN OF A SET

Linear Combination

Definition: *Let V (F) be a vector space. If $\alpha_1, \alpha_2, \ldots, \alpha_n \in V$, then any vector*

$$\alpha = a_1\alpha_1 + a_2\alpha_2 + \ldots + a_n\alpha_n \text{ where } \alpha_1, \alpha_2, \ldots, \alpha_n.$$

is called a linear combination of the vector $\alpha_1, \alpha_2, \ldots, \alpha_n$.

Linear Span

Definition: *Let V (F) be a vector space and S be any non-empty subset of V. Then the linear span of S is the set of all linear combinations of finite sets of elements of S and is denoted by L(S).* Thus we have

$L(S) = \{a_1\alpha_1 + a_2\alpha_2 + \ldots + a_n\alpha_n: \alpha_1, \alpha_2, \ldots, \alpha_n$

is any arbitrary finite subset of S and $a_1, a_2, \ldots, a_n$ is any arbitrary finite subset of F$\}$.

Theorem 1:

The linear span L(S) of any subset of S of a vector space V (F) is a subspace of V generated by S i.e. L (S) = {S}.

Proof:

Let α, β be any two elements of L(S).

Then $\alpha = a_1\alpha_1 + a_2\alpha_2 + ... + a_m\alpha_m$

and $\beta = b_1\beta_1 + b_2\beta_2 + ... + b_n\alpha_n$

where the a's and b's are elements of F and the α's and β's are element of S.

If a, b, be any two elements of F, then

$$a\alpha + b\beta = a\,(a_1\alpha_1 + a_2\alpha_2 + ... + a_m\alpha_m) + (b_1\beta_1 + b_2\beta_2 + ... + b_n\beta_n)$$
$$= a\,(a_1\alpha_1) + a\,(a_2\alpha_2) + ... + a\,(a_m\alpha_m) + b\,(b_1\beta_1) + b\,(b_2\beta_2) + ... + b\,(b_n\beta_n)$$
$$= (aa_1)\,\alpha_1 + (aa_2)\,\alpha_2 + ... + (aa_m)\,\alpha_m + (bb_1)\,\beta_1 + (bb_2)\,\beta_2 + ... + (bb_n)\,\beta_n.$$

Thus, $a\alpha + b\beta$ has been expressed as a linear combination of a finite set $\alpha_1, \alpha_2, ... \alpha_m, \beta_1, \beta_2, ..., \beta_n$ of the elements of S. Consequently $a\alpha + b\beta \in L(S)$.

Thus a, b, $\in$ F and $\alpha, \beta \in L(S) \Rightarrow a\alpha + b\beta \in L(S)$.

Hence L(S) is a subspace of V(F).

Also each element of S belongs to L(S) because if $\alpha_r \in S$, then $\alpha_r = 1\alpha_r$ and this implies that $\alpha_r \in L(S)$. Thus L(S) is a subspace of V and S is contained in L(S).

Now if W is any subspace of V containing S, then each element of L(S) must be in W because W is to be closed under vector addition and scalar multiplication. Therefore L(S) will be contained in W.

Hence L(S) = {S} *i.e.* L(S) is the smallest subspace of V containing S.

Note: If in any case we are to prove that L(S) = V, then we should prove that $V \subseteq L(S)$ because $L(S) \subseteq V$ since L(S) is a subspace of V. In order to prove that $V \subseteq L(S)$, we should prove that each element of V can be expressed as a linear combination of a finite number of elements of S. Then each element of V will also be an element of L(S) and we shall have $V \subseteq L(S)$.

Finally $V \subseteq L(S)$ and $L(S) \subseteq V \Rightarrow L(S) = V$.

Example:

The subset S = {(1, 0, 0), (0, 1, 0) (0, 0, 1)} of $V_3(F)$ generates or spans the entire vector space $V_2(F)$ i.e., L(S) = V.

Solution:

Do your self.

Hint: If (a, b, c) be any element of V, then

(a, b, c) = (1, 0, 0) + b (0, 1, 0) + c (0, 0, 1).

Thus (a, b, c) $\in$ L(S). Hence V $\subseteq$ L(S). Also L(S) $\subseteq$ V.

Hence L(S) = V.

2.7 LINEAR SUM OF TWO SUBSPACES

Definition: *Let W_1 and W_2 be two subspaces of the vector space V (F). Then the linear sum of the subspaces W_1 and W_2 denoted by $W_1 + W_2$ is the set of sums $\alpha_1 + \alpha_2$ such that $\alpha_1 \in W_1$, $\alpha_2 \in W_2$.*

Thus $\quad W_1 + W_2 = \{\alpha_1 + \alpha_2 : \alpha_1 \in W_1, \alpha_2 \in W_2\}$.

Theorem 1:

If W_1 and W_2 are subspaces of the vector space V(F), then

(i) $W_1 + W_2$ is a subspace of V (F).

(ii) $W_1 + W_2 = \{W_1 \cup W_2\}$ i.e., $L(W_1 \cup W_2) = W_1 + W_2$

Proof:

(i) Let α, β, be any two elements of $W_1 + W_2$.

Then $\alpha = \alpha_1 + \alpha_2$ and $\beta = \beta_1 + \beta_2$ where $\alpha_1, \beta_1 \in W_1$ and $\alpha_1, \beta_2 \in W_2$. If a, b, $\in$ F, we have

$$a\alpha + b\beta = a(\alpha_1 + \alpha_2) + b(\beta_1 + \beta_2) = (a\alpha_1 + b\beta_1) + (a\alpha_2 + \beta b_2)$$

Since W_1 is a subspace of V, therefore a, b, $\in$ F

and $\qquad \alpha_1, \beta_1 \in W_1 \Rightarrow a\alpha_1 + b\beta_1 \in W_1$

Similarly $\qquad a\alpha_2 + b\beta_2 \in W_2$.

Consequently $\quad a\alpha + b\beta = (a\alpha_1 + b\beta_1) + (a\alpha_2 + b\beta_2) \in W_1 + W_2$.

Thus a, b $\in$ F and $\alpha, \beta \in W_1 + W_2 \Rightarrow a\alpha + b\beta \in W_1 + W_2$.

Hence $W_1 + W_2$ is a subspace of V(F).

(ii) Since W_2 contains the zero vector, therefore if $\alpha_1 \in W_1$, then we can write $\alpha_1 = \alpha_1 + 0 \in W_1 + W_2$. Thus $W_1 \subseteq W_1 + W_2$. Similarly $W_2 \subseteq W_1 + W_2$ hence $W_1 \cup W_2 \subseteq W_1 + W_2$. Therefore $W_1 + W_2$ is a subspace of V(F) containing $W_1 \cup W_2$.

Now to prove that $W_1 + W_2 = \{W_1 \cup W_2\}$ we should prove that $W_1 + W_2 \subseteq L(W_1 \cup W_2)$ and $L(W_1 \cup W_2) \subseteq W_1 + W_2$.

Let $\alpha = \alpha_1 + \alpha_2$ be any element of $W_1 + W_2$. Then $\alpha \in W_1$ and $\alpha_2 \in W_2$. Therefore $\alpha_1, \alpha_2 \in W_1 \cup W_2$. We can write

$$\alpha_1 + \alpha_2 = 1\alpha_1 + 1\alpha_2.$$

Thus, $\alpha_1 + \alpha_2$ is a linear combination of a finite number of elements $\alpha_1, a_2 \in W_1 \cup W_2$.

Therefore $\alpha_1 + \alpha_2 \in L(W_1 \cup W_2)$.

$\therefore$ $W_1 + W_2 \subseteq L(W_1 \cup W_2)$.

Also $L(W_1 \cup W_2)$ is the smallest subspace containing $W_1 \cup W_2$ and $W_1 + W_2$ is a subspace containing $W_1 \cup W_2$. Therefore $L(W_1 \cup W_2)$ must be contained in $W_1 + W_2$. Consequently

$$L(W_1 \cup W_2) \subseteq W_1 + W_2.$$

Hence $W_1 + W_2 = L(W_1 \cup W_2) = \{W_1 \cup W_2\}$.

Theorem 2:

If S, T are subsets of V (F), then

(i) $S \subseteq T \Rightarrow L(S) \subseteq L(T)$.

(ii) $L(S \cup T) = L(S) + L(T)$.

(iii) $L(L(S)) = L(S)$.

Proof:

(i) Let $\alpha = a_1\alpha_1 + a_2\alpha_2 + \ldots a_n\alpha_n \in L(S)$, where $\{\alpha_1, \alpha_2, \ldots, \alpha_n\}$ is a finite subset of S. Since $S \subseteq T$, therefore $\{\alpha_1, \alpha_2, \ldots, \alpha_n)$ is a finite subset of T. So $\alpha \in L(T)$.

Thus $\alpha \in L(S) \Rightarrow \alpha \in L(T)$.

$\therefore L(S) \subseteq L(T)$.

(ii) Let α be any element of $L(S \cup T)$. Then

$$\alpha = a_1\alpha_1 + a_2\alpha_2 + \ldots + a_m\alpha_m + b_1\beta_1 + b_2\beta_2 + \ldots + b_p\beta_p$$

where $(\alpha_1, \alpha_2, \ldots, \alpha_m, \beta_1, \beta_2, \ldots, \beta_p)$ is a finite subset of $S \cup T$ such that $\{\alpha_1, \alpha_2, \ldots, \alpha_m)$ Í S and $\{\beta_1, \beta_2, \ldots, \beta_p\} \subset T$.

Now $a_1\alpha_1 + a_2\alpha_2 + \ldots + a_m\alpha_m \in L(S)$ and

$b_1\beta_1 + b_2\beta_2 + \ldots + b_p\beta_p \in L(T)$.

Therefore $\alpha \in L(S) + L(T)$. Consequently $L(S \cup T) \subseteq L(S) + L(T)$.

Now let γ be any element of $L(S) + L(T)$. Then $\gamma = \beta + \delta$ where $\beta \subset L(S)$ and $\delta \in L(T)$. Now β will be a linear combination of a finite number of elements of S and δ will be a linear combination of a finite number of elements of T. Therefore $\beta + \delta$ will be a linear combination of a finite number of elements of $S \subseteq T$. Thus $\beta + \delta \in L(S \cup T)$. Consequently

$$L(S) + L(T) \subseteq L(S) \cup T).$$

Hence $\quad L(S \cup T) = L(S) + L(T).$

(iii) L(L(S)) is the smallest subspace of V containing L(S). But L(S) is a subspace of V. Therefore the smallest subspace of V containing L(S) is L(S) itself.

Hence L(L(S) = L(S).

2.8 LINEAR DEPENDENCE AND LINEAR INDEPENDENCE OF VECTORS

Linear Dependence

Definition: *Let V(F) be a vector space. A finite set* $\{\alpha_1, \alpha_2 ..., \alpha_n\}$ *of vectors of V is said to be linearly dependent if there exist scalars* $a_1, a_2, ...,$ $a_n \in F$ *not all of them 0 (some of they may be zero) such that*

$$a_1\alpha_1 + a_2\alpha_2 + a_3\alpha_3 + ... + a_n\alpha_n = 0.$$

Linear Independence

Definition: *Let V(F) be a vector space. A finite set* $\{\alpha_1, \alpha_2 ..., \alpha_n\}$ *of vectors of V is said to be linearly independent if every relation of the form*

$$a_1\alpha_1 + a_2\alpha_2 + ... + a_n\alpha_n = 0,\ a_1 \in F,\ 1 \leq i \leq n$$

$$\Rightarrow a_1 = 0 \text{ for each } 1 \leq i \leq n.$$

Any **infinite set** *of vectors of V is said to be linearly independent if its every finite subset is linearly independent, otherwise it is linearly dependent.*

Example 1:

In the vector space F[x] of all polynomials over the field F the infinite set $S = \{1, x, x^2, x^3, ...\}$ *is linearly independent.*

Solution:

Let $S' = \{x^{m_1}, x^{m_2}, ..., x^{m_n}\}$ be any finite subset of S having n vectors. Here $m_1, m_2..., m_n$ are some non-negative integers. Let $a_1, a_2,..., a_n$ be scalars such that

$$a_1x^{m_1} + a_2x^{m_2} + ... + a_nx^{m_n} = 0.$$

(*i.e.*, zero polynomial) ...(1)

By the definition of equality of two polynomials we have from (1)

$$a_1 = 0,\ a_2 = 0,\ ...,\ a_n = 0.$$

Thus every finite subset of S is linearly independent.

Therefore S is linearly independent.

Example 2:

Is the vector (2, – 5, 3) in the subspace of $\mathbf{R}^3$ *spanned by the vectors (1, – 3, 2), (2, – 4, – 1), (1, – 5, 7)?*

Solution:

Let $\alpha = (2, -5, 3)$, $\alpha_1 = (1, -3, 2)$, $\alpha_2 = (2, -4, -1)$, $\alpha_3 = (1, -5, 7)$. If a can be expressed as a linear combination of the vectors $\alpha_1, \alpha_2, \alpha_3$, then it will be in the subspace of $\mathbf{R}^3$ spanned by these vectors otherwise it will not be.

Let $\alpha = a_1\alpha_1 + a_2\alpha_2 + a_3\alpha_3$ where $a_1, a_2, a_2 \in \mathbf{R}$.

Then $(2, -5, 3) = a_1 (1, -3, 2) + a_2 (2, -4, -1) + a_3 (1, -5, 7)$

$\Rightarrow (2, -5, 3) = (a_1 + 2a_2 + a_3, -3a_1 - 4a_2 - 5a_3, 2a_1 - a_2 + 7a_3)$.

$$\therefore \quad a_1 + 2a_2 + a_3 = 2 \qquad ...(1)$$

$$-3a_1 - 4a_2 - 5a_3 = -5 \qquad ...(2)$$

$$2a_1 - a_2 + 7a_3 = 3 \qquad ...(3)$$

Multiplying the equation (1) by 3 and adding to (2), we get

$$2a_2 - 2a_3 = 1 \quad \Rightarrow \quad a_2 - a_3 = \frac{1}{2}. \qquad ...(4)$$

Again multiplying the equation (1) by 2 and subtracting from (3), we get

$$-5a_2 + 5a_3 = -1 \quad \Rightarrow \quad a_2 - a_3 = 1/5. \qquad ...(5)$$

The relations (4) and (5) show that the above equations are inconsistent. Hence the vector α cannot be expressed as a linear combination of the vectors $\alpha_1, \alpha_2, \alpha_3$. Therefore α is not in the subspace of $\mathbf{R}^3$ generated by the vectors $\alpha_1, \alpha_2, \alpha_3$.

Example 3:

Show that the vectors (1, 1, 2, 4), (2, – 1, – 5, 2) (1, – 1, – 4, 0) and (2, 1, 1, 6) are linearly dependent in $\mathbf{R}^4$.

Solution:

Let $(1, 1, 2, 4) = a (2, -1, -5, 2) + b (1, -1, -4, 0) + c (1, 1, 1, 6)$.

Then $\quad 2a + b + 2c = 1 \qquad ...(1)$

$$-a - b + c = 1 \qquad \text{...(2)}$$

$$-5a - 4b + c = 2 \qquad \text{...(3)}$$

$$22a + 0b + 6c = 4. \qquad \text{...(4)}$$

Now we shall solve the simultaneous equations (1), (2), (3) and (4).

Adding (1) and (2), we get $a + 3c = 2$ which is the same equation as (4).

If we take $c = 0$, we get $a = 2$.

Putting $a = 2$ and $c = 0$ in (1), we get $b = -3$.

We see that $a = 2$, $b = 2\ 3$, $c = 0$ satisfy all the four equations (1), (2), (3) and (4).

$$\therefore (1, 1, 2, 4) = 2\,(2, -1, -5, 2) - 3\,(1, -1, -4, 0) + 0\,(2, 1, 1, 6)$$

$$\Rightarrow 1\,(1, 1, 2, 4) - 2\,(2, -1, -5, 2) + 3\,(1, -1, -4, 0) - 0\,(2, 1, 1, 6) = (0, 0, 0, 0), \qquad \text{...(1)}$$

Since in the linear relation (1) among the four given vectors the scalar coefficients $1, -2, 3, 0$ are not all zero, therefore the given vectors are linearly dependent in $\mathbf{R}^4$.

Example 4:

Show that the system of three vectors (1, 3, 2), (1, – 7, – 8), (2, 1, – 1) of $V_3(\mathbf{R})$ is linearly dependent.

Solution:

Let a, b, c be scalars *i.e.*, real numbers such that

$$a\,(1, 3, 2) + b\,(1, -7, -8) + c\,(2, 1, -1) = (0, 0, 0)$$

i.e. $\quad (a + b + 2c, 3a - 7b + c, 2a - 8b - c) = (0, 0, 0).$

Then $\quad a + b + 2c = 0 \qquad \text{...(1)}$

$$3a - 7b + c = 0, \qquad \text{...(2)}$$

and $\quad 2a - 8b - c = 0. \qquad \text{...(3)}$

If the equations (1), (2), (3) posses a non-zero solution, then the given vectors are linearly dependent.

Adding (2) and (3), we get

$$5a - 15b = 0 \qquad \Rightarrow a - 3b = 0. \qquad \text{...(4)}$$

Multiplying (3) by (2) and adding to (1), we get

$$5a - 15b = 0$$

$\Rightarrow \quad a - 3b = 0.$ ...(5)

The equations (4) and (5) are the same and give $a = 3b$.

Putting $a = 3b$ in (1), we get $\quad 2c + 4b = 0 \Rightarrow c = -2b$.

If we take $b = 1$, we get $a = 3$, $c = -2$.

Thus $a = 3$, $b = 1$, $c = -2$ is non-zero solution of the equations (1), (2) and (3). Hence the given set of vectors is linearly dependent.

Alternative Method. The coefficient matrix A of the system of equations (1), (2) and (3) is

$$A = \begin{bmatrix} 1 & 1 & 2 \\ 3 & -7 & 1 \\ 2 & -8 & -1 \end{bmatrix}.$$

We have $\det A = \begin{vmatrix} 1 & 1 & 2 \\ 3 & -7 & 1 \\ 2 & -8 & -1 \end{vmatrix}$

$$= \begin{vmatrix} 1 & 0 & 0 \\ 3 & -10 & -5 \\ 2 & -10 & -1 \end{vmatrix}, \text{ by } C_2 - C_1 \text{ and } C_3 - 2C_1$$

$= 50 - 50 = 0.$

$\therefore$ rank $A < 3$ *i.e.*, rank $A <$ the number of unknowns a, b, c in the equations (1), (2) and (3).

Therefore the equations (1), (2) and (3) must posses a non-zero solution. Hence the given vectors are linearly dependent.

Example 5:

Find whether the vectors $2x^3 + x^2 + x + 1$, $x^3 + 3x^2 + x - 2$ and $x^3 + 2x^2 - x + 3$ of **R** *[x], the vector space of all polynomials over the real number field, are linearly independent or not.*

Solution:

The zero vector of the vector space **R** [x] is the zero polynomial.

Let a, b, c be scalars (*i.e.*, real numbers) such that

$a (2x^3 + x^2 + x + 1) + b (x^2 + 3x^2 + x - 2) + c (x^3 + 2x^2 - x + 3) = 0$

i.e., zero polynomial.

Then $\quad (2a+b+c)x^2+(a+3b+2c)x^2+(a+b-c)x+a-2b+3c=0$ ...(1)

Equating the coefficients of like powers of x on both sides of (1), we get

$$\begin{aligned} 2a + b + c &= 0, \\ a + 3b + 2c &= 0, \\ a + b - c &= 0, \\ \text{and} \quad a - 2b + 3c &= 0. \end{aligned} \qquad \text{...(2)}$$

The coefficients matrix A of the system of equations (2) is

$$A = \begin{bmatrix} 2 & 1 & 1 \\ 1 & 3 & 2 \\ 1 & 1 & -1 \\ 1 & -2 & 3 \end{bmatrix}$$

$$\sim \begin{bmatrix} 1 & 3 & 2 \\ 2 & 1 & 1 \\ 1 & 1 & -1 \\ 1 & -2 & 3 \end{bmatrix}, \text{ by } R_1 \leftrightarrow R_2$$

$$\sim \begin{bmatrix} 1 & 3 & 2 \\ 0 & -5 & -3 \\ 0 & -2 & -3 \\ 0 & -5 & 1 \end{bmatrix},$$

$$\sim \begin{bmatrix} 1 & 3 & 2 \\ 0 & 1 & 3/5 \\ 0 & -2 & -3 \\ 0 & -5 & 1 \end{bmatrix}, \text{ by } R_2 \to -\frac{1}{5}R_2$$

$$\sim \begin{bmatrix} 1 & 3 & 2 \\ 0 & 1 & 3/5 \\ 0 & 0 & -9/5 \\ 0 & 0 & 4 \end{bmatrix}, \text{ by } R_3 \to R_3 + 2R_2,\ R_4 \to R_4 + 5R_2$$

$$\sim \begin{bmatrix} 1 & 3 & 2 \\ 0 & 1 & 3/5 \\ 0 & 0 & 1 \\ 0 & 0 & 4 \end{bmatrix}, \text{ by } R_3 \to -5/9R_3$$

$$\sim \begin{bmatrix} 1 & 3 & 2 \\ 0 & 1 & 3/5 \\ 0 & 0 & 1 \\ 0 & 0 & 0 \end{bmatrix}, \text{ by } R_4 \to R_4 - 4R_3$$

which is in echelon form.

$\therefore$ rank A = number of non-zero row in the its echelon form

= 3 = number of unknowns a, b, c in the system of equations (2).

Hence the system of equations (2) has the only solution

$$a = 0,\ b = 0,\ c = 0.$$

$\therefore$ the given set of vectors is linearly independent.

Example 6:

Determine whether the following set of vectors in V_3 (Q) is linearly dependent or independent, Q being the field of rational numbers:

$$\{(-1, 2, 1,), (3, 1, -2)\}$$

Solution:

Let a, b be scalars (*i.e.*, a, b, Î Q) such that

$$a(-1, 2, 1,) + b(3, 1, -2) = (0, 0, 0)$$

i.e., $(-a + 3b, 2a + b, a - 2b) = (0, 0, 0).$

Then
$$\begin{aligned} -a + 3b &= 0, \\ 2a + b &= 0, \\ a - 2b &= 0, \end{aligned} \qquad \text{...(1)}$$

The coefficient matrix A of the system of equations (1) is

$$A = \begin{bmatrix} -1 & 3 \\ 2 & 1 \\ 1 & -2 \end{bmatrix}.$$

We have $\begin{vmatrix} -1 & 3 \\ 2 & 1 \end{vmatrix} = -1 - 6 = -7 \neq 0.$

Thus, there exists a 2-rowed minor of the matrix A which is not zero. Also the matrix A can have no minor of order greater than 2.

$\therefore$ rank A = 2 = the number of unknowns a and b.

Therefore the equations (1) have the only solution a = 0, b = 0. Hence the given get of vectors is linearly independent.

Note: If we do not want to use the concept of the rank of a matrix to discuss the solutions of the system of equations (1), we can directly say that solving the system of equations (1) we find that the only solution of the system of equations (1) is a = 0, b = 0.

Example 7:

Find a linearly independent subset T of the set

$S = \{\alpha_1, \alpha_2, \alpha_3, \alpha_4)$

where $\alpha_1 = (1, 2, -1)$, $\alpha_2 = (-3, -6, 3)$,

$\alpha_3 = (2, 1, 3)$, $\alpha_4 = (8, 7, 7) \in \mathbf{R}^3$

which spans the same space as S.

Solution:

First we observe that $\alpha_2 = -3\alpha_1$ so that the vectors α_1 and α_2 are linearly dependent.

$\therefore$ If $S_1 = \{\alpha_1, \alpha_3, \alpha_4\}$, then the subspace of $\mathbf{R}^3$ spanned by S_1 is the same as that spanned by S.

Now there exists no real number c such that $\alpha_3 = c\alpha_1$. Therefore the vectors α_1 and α_3 are linearly independent.

Let us now see whether the vector α_4 lies in the subspace of $\mathbf{R}^3$ spanned by the vectors α_1 and α_3 or not.

Let $\alpha_4 = a\alpha_1 + b\alpha_3$, where $a, b \in \mathbf{R}$.

Then $(8, 7, 7) = a\,(1, 2, -1) + b\,(2, 1, 3)$.

$\therefore \quad a + 2b = 8.$

$2a + b = 7,$

and $\quad -a + 3b = 7.$

Solving the first two of these three equations, we get a = 2, b = 3. These values of a and b also satisfy the third equation.

$\therefore \quad \alpha_4 = 2\alpha_1 + 3\alpha_3.$

Thus, the vector α_4 has been expressed as a linear combination of α_1 and α_3 so that the subspace of $\mathbf{R}^3$ spanned by the vectors α_1, α_3, and α_4 is the same as that spanned by the vectors α_1 and α_3.

Hence $T = \{\alpha_1, \alpha_3\}$ is a linearly independent subset of S which spans the same subspace of $\mathbf{R}^3$ as is spanned by S.

Example 8:

Prove that a set of vectors which contains the zero vector is linearly dependent.

Solution:

Let $S = \{\alpha_1, \alpha_2, ..., \alpha_n\}$ be a set of vectors of the vector space V(F).

Let α_r be equal to zero vector where $1 \leq r \leq n$.

To show that the set S of vectors in linearly dependent.

Obviously,

$0\alpha_1 + 0\alpha_2 + ... + a\alpha_r + 0\alpha_{r+1} + ... + 0\alpha_n = 0$ *i.e.*, zero vector ...(1)

for any non-zero scalar a *i.e.*, for any non-zero element a in the field F.

Since in the linear relation (1) among the vectors $\alpha_1, \alpha_2, ..., \alpha_n$, the scalar coefficient a is not zero, therefore the vectors $\alpha_1, \alpha_2, ..., \alpha_n$ are linearly dependent.

Example 9:

Show that the set {1, x, x (1 – x)} is a linearly independent set of vectors in the space of all polynomials over the real number field.

Solution:

The zero vector of the vector space of all polynomials over the real number field is the zero polynomial.

Let a, b, c be scalars (*i.e.*, real numbers) such that

$$a(1) + bx + c\,[x\,(1 - x)] = 0 \text{ i.e., zero polynomial.}$$

We have $a(1) + bx + c\,(x - x^2) = 0$

$$\Rightarrow \quad a + (b + c)\,x - cx^2 = 0. \qquad ...(1)$$

Now two polynomials in x are said to be equal if the coefficients of like powers of x on both sides are equal. So by the definition of the equality of two polynomials, we have from (1).

$$a = 0,\ b + c = 0,\ - c = 0$$

$$\Rightarrow \quad c = 0,\ b = 0,\ a = 0.$$

Thus $\quad a(1) + bx + c\,[x\,(1 - x)] = 0$

$$\Rightarrow \quad a = 0,\ b = 0,\ c = 0.$$

$\therefore$ the vectors 1, x, x (1 – x) are linearly independent over the field of real numbers.

Example 10:

Show that the three vectors (1, 1, – 1), (2, – 3, 5) and (–2, 1, 4) of $\mathbf{R}^3$ *are linearly independent.*

Solution:

Let a, b, c be scalars *i.e.*, real numbers such that

a $\quad (1, 1, -1) + b\,(2, -3, 5) + c\,(-2, 1, 4) = (0, 0, 0)$

i.e., $\quad (a + 2b - 2c, a - 3b + c, -a + 5b + 4c) = (0, 0, 0)$

i.e.
$$a + 2b - 2c = 0 \quad ...(1)$$
$$a - 3b + c = 0 \quad ...(2)$$
$$-a + 5b + 4c = 0 \quad ...(3)$$

Now we shall solve the simultaneous equations (1), (2) and (3).

Multiplying (2) by 2 and adding to (1), we get

$$3a - 4b = 0 \quad ...(4)$$

Again multiplying (1) by 2 and adding to (3), we get

$$a + 9b = 0$$

Multiplying (5) by 3 and subtracting from (4), we get

$$-31b = 0 \text{ or } b = 0.$$

Putting $b = 0$ in (5), we get $a = 0$.

Now putting $a = 0$, $b = 0$ in (1), we get $c = 0$.

Thus, $a = 0$, $b = 0$, $c = 0$ is the only solution of the equations (1), (2) and (3).

$\therefore \quad a\,(1, 1, -1) + b\,(2, -3, 5) + c\,(-2, 1, 4) = (0, 0, 0)$

$\Rightarrow \quad a = 0, b = 0, c = 0.$

Hence the vectors $(1, 1, -1)$, $(2, -3, 5)$, $(-2, 1, 4)$ of $\mathbf{R}^3$ are linearly independent.

Example 11:

If α, β, γ are linearly independent vectors of $V(F)$ where F is the field of complex numbers, then so also are $\alpha + \beta$, $\beta + \gamma$, $\gamma + \alpha$.

Solution:

Let a, b, c be scalars such that

$$a\,(\alpha + \beta) + b\,(\beta + \gamma) + (\gamma + \alpha) = 0$$

i.e., $\quad (a + c)\,\alpha + (a + b) + \beta\,(b + c) + \gamma = 0. \quad ...(1)$

But α, β, γ are linearly independent. Therefore (1) implies

$$a + 0b + c = 0,\ a + b + 0c = 0,\ 0a + b + c = 0.$$

The coefficient matrix A of these equations is

$$A = \begin{bmatrix} 1 & 0 & 1 \\ 1 & 1 & 0 \\ 0 & 1 & 1 \end{bmatrix}.$$

We have rank A = 3 *i.e.*, the number of unknowns a, b, c. Therefore a = 0, b = 0, c = 0 is the only solution of the given equations.

Hence $\alpha + \beta$, $\beta + \gamma$, $\gamma + \alpha$ are also linearly independent.

Example 12:

If α, β, γ are linearly independent vectors of V(F) where F is the field of complex numbers, then so also are

$$\alpha + \beta,\ \alpha - \beta,\ \alpha - 2\beta + \gamma.$$

Solution:

Let a, b, c be scalars such that

$$(a + \beta) + b\,(\alpha - \beta) + c\,(\alpha - 2\beta + \gamma) = 0 \qquad ...(1)$$

i.e., $$(a + b + c)\,\alpha + (a - b - 2c)\,\beta + c\gamma = 0. \qquad ...(2)$$

But α, β, γ are linearly independent. Therefore (2) implies

$$a + b + c = 0,\ a - b - 2c = 0,\ c = 0.$$

The only solution of these equations is c = 0, a = 0, b = 0.

Thus (1) implies a = 0, b = 0, c = 0. Therefore the vectors $\alpha + \beta$, $\alpha - \beta$, $\alpha - 2\beta + \gamma$ are linearly independent.

Example 13:

A system consisting of a single non-zero vector is always linearly independent.

Solution:

Let S = $\{\alpha\}$ be a subset of a vector space V and let α be not equal to zero vector. If a is any scalar, then

$$a\alpha = 0$$

$$\Rightarrow \quad a = 0 \qquad \text{[Since a is not zero vector]}$$

$\therefore$ the set S is linearly independent.

Example 14:

If the set $S - \{u_1, u_2, ..., u_n\}$

of vectors of V(F) is linearly independent, then of the vectors $\alpha_1, \alpha_2, ..., \alpha_n$ can be zero vector.

Solution:

Let α_r be equal to zero vector where $1 \leq r \leq n$. Then

$$0\alpha_1 + 0\alpha_2 + \ldots + a\alpha_r + 0\alpha_{r+1} + \ldots + 0\alpha_n = 0$$

for any $a \neq 0$ in F.

Since $a \neq 0$, therefore from this relation we conclude that S is linearly dependent. Thus we get a contradiction because it is given that S is linearly independent. Hence none of the vectors $\alpha_1, \alpha_2 ..., \alpha_n$ can be zero vector. We also conclude that *a set of vectors which contains the zero vector is necessarily linearly dependent.*

Example 15:

Every superset of a linearly dependent set of vectors is linearly dependent.

Solution:

Let $S = \{\alpha_1, \alpha_2, ..., \alpha_n\}$ be a linearly dependent set of vectors. Then there exist scalars $a_1, a_2, ..., a_n$ not all zero such that

$$a_1\alpha_1 + a_2\alpha_2 + \ldots + a_n\alpha_n = 0. \qquad \ldots(1)$$

Now let $S' = \{\alpha_1, \alpha_2, \ldots, \alpha_n, \beta_1, \beta_2, \ldots, \beta_m)$ be a superset of S. Then we have from (1)

$$a_1\alpha_1 + a_2\alpha_2 + \ldots + a_n\alpha_n + 0b_1 + 0\beta_2 + \ldots + 0\beta_m = 0. \qquad \ldots(2)$$

Since in the relation (2) the scalar coefficients are not all 0, therefore S' is linearly dependent.

From this we also conclude that *any subset of a linearly independent set of vectors is also linearly independent.*

Example 16:

Show that

$S = \{(1, 2, 4), (1, 0, 0), (0, 1, 0), (0, 0, 1)\}$

*is a linearly dependent subset of the vector space V_3(**R**) where **R** is the field of real numbers.*

Solution:

We have

$1\ (1, 2, 4) + (-1)\ (1, 0, 0) + (-2)\ (0, 1, 0) + (-4)\ (0, 0, 1)$

$= (1, 2, 4) + (-1, 0, 0) + (0, -2, 0) + (0, 0, -4)$

$= (0, 0, 0)$ *i.e.*, zero vector.

Since in this relation the coefficients $1, -1, -2, -4$ are not all zero therefore the given system S is linearly dependent.

Example 17:

Prove that if two vectors are linearly dependent, one of them is a scalar multiple of the other.

Solution:

Let α, β be two linearly dependent vectors of the vector space V. Then $\exists$ scalars a, b not both zero, such that

$$a\alpha + b\beta + 0.$$

If $a \neq 0$, then we get

$$a\alpha = -b\beta \Rightarrow a^{-1}(a\alpha) = a^{-1}(-b\beta)$$

$$\Rightarrow \quad (a^{-1}a)\alpha = [a^{-1}(-b)]\beta \Rightarrow 1\alpha = (-a^{-1}b)\beta$$

$$\Rightarrow \quad \alpha = -\frac{b}{a}\beta \Rightarrow \alpha \text{ is a scalar multiple of } \beta.$$

If $b \neq 0$, then we get

$$b\beta = -a\alpha$$

$$\Rightarrow \quad \beta = \left(-\frac{a}{b}\right)\alpha \Rightarrow \beta \text{ is a scalar multiple of } \alpha.$$

Thus one of the vectors α and β is a scalar multiple of the other.

Example 18:

In the vector $V_n(F)$, the system of n vectors

$$e_1 = (1, 0, 0, ..., 0),$$
$$e_2 = (0, 1, ..., 0, 0),$$
$$e_n = (0, 0, ..., 0, 1)$$

is linearly independent where 1 denotes the unity of the field F.

Solution:

If $a_1, a_2, a_3, ..., a_n$ be any scalars, then

$$a.e_1 + a_2e_2 + ... + a_ne_n = 0$$

$$\Rightarrow a_1(1, 0, 0, ..., 0) + a_2(0, 1, 0, ..., 0) + ... + a_n.(0, 0, ..., 1) = 0$$

$$\Rightarrow (a_1, a_2, ..., a_n) = (0, 0, ..., 0) \Rightarrow a_1 = 0, a_2 = 0, ..., a_n = 0.$$

Therefore the given set of n vectors is linearly independent.

In particular {(1, 0, 0), (0, 1, 0), (0, 0, 1)} is a linearly independent subset of $V_3(F)$.

Example 19:

Let α_1, α_2, α_3 be vectors of V(F), a, b, ∈ F. Show that the set $\{\alpha_1, \alpha_2, \alpha_3\}$ is linearly dependent if the set $\{\alpha_1 + a\alpha_2 + b\alpha_3, \alpha_2, \alpha_3\}$ is linearly dependent.

Solution:

Since the set $\{\alpha_1 + a\alpha_2 + b\alpha_3, \alpha_2, \alpha_3\}$ is linearly dependent therefore exist scalars x, y, z not all zero such that

$$x(\alpha_1 + a\alpha_2 + b\alpha_3) + y\alpha_2 + z\alpha_3 = 0$$

i.e.

$$x\alpha_1 + (xa + y)\alpha_2 + (xb + z)\alpha_3 = 0. \qquad ...(1)$$

If in the relation (1), the coefficients x, xa + y, xb + z are not all zero, then the set $\{\alpha_1, \alpha_2, \alpha_3\}$ will also be linearly dependent.

If $x \neq 0$, then the problem is at once solved whatever y and z may be. However if x = 0, then at least one of y and z is not zero. Therefore at least one of xa + y and xb + z will not be zero since when x = 0 then xa + y and xb + z reduce to y and z respectively.

Hence in the relation (1) the scalar coefficients of α_1, α_2, α_3 are not all zero. Therefore the set $\{\alpha_1, \alpha_2, \alpha_3\}$ is also linearly dependent.

Example 20:

*In $V_3(\mathbf{R})$, where **R** is the field of real numbers, examine each of the following sets of vectors for linear dependence:*

(i) {(2, 1, 2), (8, 4, 8)}.

(ii) {(1, 2, 0), (0, 3, 1), (– 1, 0, 1)}

(iii) {(– 1, 2, 1), (3, 0, – 1), (– 5, 4, 3)}.

(iv) {(2, 3, 5), (4, 9, 25)}.

(v) (1, 2, 1), (3, 1, 5), (3, – 4, 7).

Solution:

(i) We have

4 (2, 1, 2) + (– 1) (8, 4, 8) = (8, 4, 8) + (– 8, – 4, – 8)

= (0, 0, 0) *i.e.*, the zero vector.

Since in this relation the scalar coefficients 4, – 1 are not both zero, therefore the given set is linearly dependent.

(ii) Let a, b, c be scalars *i.e.*, real numbers such that

a (1, 2, 0) + b (0, 3, 1) + c (– 1, 0, 1) = (0, 0, 0)

i.e., $(a - c, 2a + 3b, b + c) = (0, 0, 0)$

i.e., $a + 0b - c, 2a + 3b + 0c = 0, 0a + b + c = 0.$

These equations will have a non-zero solution *i.e.*, a solution in which a, b, c are not all zero if the rank of the coefficient matrix is less than three *i.e.*, the number of unknowns a, b, c. If the rank is 3, then the zero solution a = 0, b = 0, c = 0 will be the only solution.

Coefficient matrix $A = \begin{bmatrix} 1 & 0 & -1 \\ 2 & 3 & 0 \\ 0 & 1 & 1 \end{bmatrix}.$

We have $|A| = 1\,(3 - 0) - 2\,(0 + 1) = 1 \neq 0.$

$\therefore$ Rank A = 3. Hence a = 0, b = 0, c = 0 is the only solution. Therefore the given system is linearly independent.

(iii) Let a, b, c be scalars such that

$$a\,(-1, 2, 1) + b\,(3, 0, -1) + c\,(-5, 4, 3) = (0, 0, 0)$$

i.e., $(-a + 3b - 5c, 2a + 0b + 4c, a - b + 3c) = (0, 0, 0)$

i.e., $-a + 3b - 5c = 0, 2a + 0b + 4c = 0, a - b + 3c = 0.$

The coefficient matrix A of these equations is

$$A = \begin{bmatrix} -1 & 3 & -5 \\ 2 & 0 & 4 \\ 1 & -1 & 3 \end{bmatrix}.$$

We have $|A| = -(0 + 4) - 2\,(9 - 5) + 1\,(12 - 0) = 0.$

$\therefore$ Rank A is < 3 *i.e.*, the number of unknowns a, b, c. Therefore the given system of equations will possess a non-zero solution. For example a = –2, b = 1, c = 1, is a non-zero solution. Hence the given system of vectors is linearly dependent.

(iv) Let a, b, be scalars *i.e.*, real numbers such that

$$a\,(2, 3, 5) + b\,(4, 9, 25) = (0, 0, 0)$$

i.e., $(2a + 4b, 3a + 9b, 5a + 25b) = (0, 0, 0)$

i.e., $2a + 4b = 0, 3a + 9b = 0, 5a + 25b = 0.$

The coefficients matrix A of these equations is

$$\begin{bmatrix} 2 & 4 \\ 3 & 9 \\ 5 & 25 \end{bmatrix}.$$

Obviously rank A = 2 *i.e.*, equal to the number of unknowns a and b.

Therefore these equations have the only solution $a = 0$, $b = 0$. Hence the given set of vectors is linearly independent.

(v) Let a, b, c be scalars *i.e.*, real numbers such that

$$a(1, 2, 1) + b(3, 1, 5) + c(3, -4, 7) = (0, 0, 0)$$

i.e., $$(a + 3b + 3c, 2a + b - 4c, a + 5b + 7c) = (0, 0, 0)$$

i.e., $$a + 3b + 3c = 0, \quad ...(1)$$

$$2a + b - 4c = 0. \quad ...(2)$$

$$a + 5b + 7c = 0. \quad ...(3)$$

Multiplying (1) by 2, we get

$$2a + 6b + 6c = 0. \quad ...(4)$$

Subtracting (4) from (2), we get

$$-5b - 10c = 0$$

$$\Rightarrow \quad b + 2c = 0. \quad ...(5)$$

Again subtracting (3) from (1), we get

$$-2b - 4c = 0,$$

$$\Rightarrow \quad b + 2c = 0. \quad ...(6)$$

The equations (5) and (6) are the same and give $b = -2c$,

Putting $b - 2c$ in (1), we get $a = 3c$. If we take $c = 1$, we get $b = -2$ and $a = 3$. Thus $b = 3$, $b = -2$, $c = 1$ is a non-zero solution of the equations (1), (2) and (3). Hence the given set of vectors is linearly dependent.

Example 21:

If F is the field of complex numbers, prove that the vectors (a_1, a_2) and (b_1, b_2) in $V_2(F)$ are linearly dependent iff

$$a_1b_2 - a_2b_2 = 0.$$

Solution:

Let $x, y \in F$. Then

$$x(a_1, a_2) + y(b_1, b_2) = (0, 0)$$

$$\Rightarrow (xa_1 + by_1, xa_2 + yb_2) = (0, 0).$$

Therefore $a_1x + b_1y = 0$

and $a_2x + b_2y = 0.$

The necessary and sufficient condition for these equations to possess a non-zero solution is that

$$\begin{vmatrix} a_1 & b_1 \\ a_2 & b_2 \end{vmatrix} = 0 \text{ i.e., } a_1b_2 - a_2b_1 = 0.$$

Hence the given system is linearly dependent iff

$$a_1b_2 - a_2b_1 = 0.$$

Example 22:

If α_1 and α_2 are vectors of V(F), and a, b $\in$ F, show that the set $\{\alpha_1, \alpha_2, a\alpha_1 + b\alpha_2)$ is linearly dependent.

Solution:

We have

$(-a)\ \alpha_1 + (-b)\ \alpha_2 + 1\ (a\alpha_1 + b\alpha_2)$

$= (-a + a)\ \alpha_1 + (-b + b)\ \alpha_2 = 0\alpha_1 + 0\alpha_2 = 0$ *i.e.*, zero vector.

In the above linear combination the scalar coefficient 1 ' 0. Therefore whatever may be the scalars – a and – b, the given set of vectors is linearly dependent.

Example 23:

In the vector space $\mathbf{R}^3$, *let* $\alpha = (1, 2, 1), \beta = (3, 1, 5), \gamma = (3, -4, 7)$. *Show that the subspaces spanned by S $\{\alpha, \beta\}$ and T = $\{\alpha, \beta, \gamma)$ are the same.*

Solution:

First we shall show that the vector g can be expressed as a linear combination of the vectors a and b. Let

$$(3, -4, 7) = a\ (1, 2, 1) + b\ (3, 1, 5).$$

Then a + 3b – 3, 2a + b = – 4, a + 5b = 7. Solving the first two equations we get a – – 3, b = 2 and these satisfy the third equation also. Therefore we can write $\gamma = -3\alpha + 2\beta$.

Now $\quad S \subseteq T \Rightarrow L(S) \subseteq L(T)$.

Further let $\delta \in L(T)$. Then δ can be expressed as a linear combination of the vectors, α, β and γ. In this linear combination the vector γ can be replaced by $-3\alpha + 2\beta$. Thus, δ can be expressed as a linear combination of the vectors α and β. Therefore d $\in$ L(S). Thus d $\in$ L(T) $\Rightarrow$ d $\in$ L(S). Therefore $L(T) \subseteq L(S)$.

Hence $\quad L(T) = L(S)$.

Example 24:

Show that the set {1, x 1 + x + x^2} is a linearly independent set of vectors in the vector space of all polynomials over the real number field.

Solution:

Let a, b, c be scalars (real numbers) such that

$$a(1) + bx + c(1 + x + x^2) = 0. \text{ We have}$$

$$a(1) + bx + c(1 + x + x^2) = 0$$

$$\Rightarrow \quad (a + c) + (b + c)x + cx^2 = 0$$

$$\Rightarrow \quad a + c = 0,\ b + c = 0,\ c = 0$$

$$\Rightarrow \quad c = 0,\ b = 0,\ a = 0.$$

$\therefore$ the vectors 1, x, $1 + x + x^2$ are linearly independent over the field of real numbers.

Example 25:

Show that the vectors (1, 1, 0, 0), (0, 1, – 1, 0), (0, 0, 0, 3) in $\mathbf{R}^4$ *are linearly independent.*

Solution:

Let a, b, c be scalars *i.e.*, real numbers such that

$$a(1, 1, 0, 0) + b(0, 1, -1, 0) + c(0, 0, 0, 3) = (0, 0, 0, 0) \quad \text{...(1)}$$

Then

$$a + 0b + 0c = 0,$$

$$a + b + 0c = 0,$$

$$0a - b + 0c = 0,$$

$$0a + 0b + 3c = 0.$$

The only solution of the above equations is

$$a = 0,\ b = 0,\ c = 0.$$

Thus the linear relation (1) among the three given vectors is possible only if a = 0, b = 0, c = 0.

Hence the three given vectors in $\mathbf{R}^4$ are linearly independent.

Example 26:

Is the vector (3, – 1, 0, – 1) in the subspace of $\mathbf{R}^4$ *spanned by the vectors (2, – 1, 3, 2), (– 1, 1, 1, – 3) and (1, 1, 9, – 5)?*

Solution:

Let $\alpha = (3, -1, 0, -1)$, $\alpha_1 = (2, -1, 3, 2)$,

$\alpha_2 = (-1, 1, 1, -3)$, $\alpha_3 = (1, 1, 9, -5)$.

If a can be expressed as a linear combination of the vectors $\alpha_1, \alpha_2, \alpha_3$, then it will be in the subspace of $\mathbf{R}^4$ spanned by these vectors otherwise it will not be.

Let $\alpha = a\alpha_1 + b\alpha_2 + c\alpha_3$ where a, b, c $\in \mathbf{R}$,

Then $(3, -1, 0, -1) = a(2, -1, 3, 2) + b(-1, 1, 1, -3)$
$+ c(1, 1, 9, -5)$.

$\therefore \quad 2a - b + c = 3, \quad \text{...(1)}$

$-a + b + c = -1, \quad \text{...(2)}$

$3a + b + 9c = 0, \quad \text{...(3)}$

and $\quad 2a - 3b - 5c = -1. \quad \text{...(4)}$

Adding the equations (1) and (2), we get

$a + 2c = 2$

Again adding the equations (1) and (3) we get

$5a + 10c = 3 \quad \text{...(6)}$

Multiplying the equation (5) by 5, we get

$5a + 10c = 10. \quad \text{...(7)}$

The relations (6) and (7) show that the equations (1), (2), (3) and (4) are inconsistent *i.e.*, do not posses a common solution. Hence the vector α cannot be expressed as a linear combination of the vectors $\alpha_1, \alpha_2, \alpha_3$. Therefore α is not in the subspace of $\mathbf{R}^4$ generated by the vectors $\alpha_1, \alpha_2, \alpha_3$.

2.9 SOME THEOREMS OF LINEAR DEPENDENCE AND LINEAR INDEPENDENCE

Theorem 1:

If in a vector space V(F), a vector β is a linear combination of the set of vectors $\alpha_1, \alpha_2, \alpha_3, \ldots, \alpha_n$, then the set of vectors $\beta, \alpha_1 \alpha_1, \ldots, \alpha_n$ is linearly dependent.

Proof:

Since b is a linear combination of $\alpha_1, \alpha_2, \ldots, \alpha_n$, therefore there exist scalars $a_1, a_2, \ldots, a_n$ such that

$$\beta = a_1\alpha_1 + a_2\alpha_2 + \dots a_n\alpha_n$$

$\Rightarrow \qquad \beta_1 - a_1\alpha_1 - a_2\alpha_3 - \dots - a_n\alpha_n = 0.$ ...(1)

In the relation (1) the scalar coefficient of β is 1 which is $\neq 0$. Hence in the relation (1) not all the scalar coefficients are 0. Therefore the set of vectors $\beta, \alpha_1, \alpha_2, \alpha_2, \dots, \alpha_n$ is linearly dependent.

Theorem 2:

Let V(F) be a vector space. If $\alpha_1, \alpha_2, \dots, \alpha_n$ are non-zero vector $\in V$ then either they are linearly independent or some α_k, $2 \leq k \leq n$, is a linear combination of the preceding ones, $\alpha_1, \alpha_2, \dots, \alpha_{k-1}$.

Proof:

If $\alpha_1, \alpha_2, \alpha_3, \dots, \alpha_n$ are linearly independent, we are nothing to prove. So let $\alpha_1, \alpha_2, \dots, \alpha_n$ be linearly dependent. Then there exists a relation of the form

$$a_1\alpha_1 + a_2\alpha_2 + \dots + a_n\alpha_n = 0 \qquad \dots(1)$$

where not all the scalar coefficients $a_1, a_2, \dots, a_n$ are 0. Let k be the largest integer for which $a_k \neq 0$ *i.e.*, $a_{k+1} = 0, a_{k+2} = 0, \dots, a_n = 0$ and $a_k \neq 0$. There is no harm in this assumption because at the most if $a_n \neq 0$ then $k = n$.

Also $2 \leq k$. Because if $a_2 = 0, a_3 = 0, \dots, a_n = 0$, then $a_1\alpha_1 = 0$ and $\alpha_1 \neq 0 \Rightarrow a_1 = 0$. This contradicts the fact that not all the a's are 0.

Now the relation (1) reduces to

$$a_1\alpha_1 + a_2\alpha_2 + \dots + a_k\alpha_k = 0, \text{ where } a_k \neq 0$$

$$\Rightarrow \quad a_k\alpha_k = -a_1\alpha_1 - a_2\alpha_2 - \dots, -a_{k-1}\alpha_{k-1}$$

$$\Rightarrow \quad a_k^{-1}(a_k\alpha_k) = a_k^{-1}(-a_1\alpha_1 - a_2\alpha_2 - \dots - a_{k-1}\alpha_{k-1})$$

$$\Rightarrow \quad \alpha_k = (-a_k^{-1}a_1)\alpha_1 + (-a_k^{-1}a_1)\alpha_2 + \dots + (-a_k^{-1}a_{k-1})\alpha_{k-1}.$$

Thus a_k is a linear combination of its preceding vectors.

Theorem 3:

The set of non-zero vectors $\alpha_1, \alpha_2, \dots, \alpha_n$ of V(F) is linearly dependent if some a_k, $2 \leq k \leq n$, is a linear combination of the proceeding ones.

Proof:

If some α_k, $2 \leq k \leq n$ is a linear combination of the proceeding ones, $\alpha_1 \alpha_2, \dots, \alpha_{k-1}$, then $\exists$ scalars $a_1, a_2, \dots, a_{k-1}$ such that $\alpha_k = a_1\alpha_1 + a_2\alpha_2 + \dots + a_{k-1}\alpha_{k-1}$

$\Rightarrow \quad 1\alpha_k - a_1\alpha_1 - a_2\alpha_2 - ... - ak_{k-1}\ \alpha_{k-1} = 0$ ...(1)

$\Rightarrow$ the set $\{\alpha_1, \alpha_2,..., \alpha_k\}$ is linearly dependent because in the linear combination (1) the scalar coefficient $1 \neq 0$.

Hence, the set $\{\alpha_1, \alpha_2,.., \alpha_n\}$ of which $\{\alpha_1, \alpha_2,..., \alpha_k\}$ is a subset must be linearly dependent.

2.10 BASIS OF A VECTOR SPACE

Definition: *A subset S of a vector space V(F) is said to be a basis of V(F), if*

(i) S consists of linearly independent vectors,

(ii) S generates V(F) i.e., L(S) = V i.e., each vector in V is a linear combination of a finite number of elements of S.

Example 1:

A system S consisting of n vectors,

$e_1 = (1, 0, 0,..., 0)$, $e_2 = (0, 1, 0,..., 0)$,..., $e_n = (0, 0,..., 0, 1)$ is a basis of V_n (F).

Solution:

First we should show that S is a linearly independent set of vectors. We have proved it in one of the previous examples.

Now we should prove that $L(S) = V_n$ (F). We have always $L(S) \subseteq V_n(F)$. So we should prove that $V_n(F) \subseteq L$ (S) *i.e.*, each vector in $V_n(F)$ is a linear combination of elements of S.

Let $\alpha = (a_1, a_3..., a_n)$ be any vector in $V_n(F)$. We can write

$(a_1, a_2,..., a_n) = a_1 (1, 0,..., 0) + a_2 (0, 1,..., 0) + ... + a_n (0, 0,..., 0, 1)$

i.e., $\alpha = a_1e_1 + a_2e_2 + ... + a_ne_n$.

Hence S is a basis of $V_n(F)$. We shall this particular basis the standard basis of $V_n(F)$.

Note: The set {(1, 0), (0, 1)} is a basis of $V_2(F)$. The set {(1, 0, 0), (0, 1, 0), (0, 0, 1)} is a basis of V_3 (F). As a particular case a basis of F(F) is the set consisting of only the unit element of F.

Example 2:

Show that the infinite set

$$S = \{1, x, x^3,..., x^n,...\}$$

is a basis of the vector space F [x] of polynomials over the field F.

Solution:

First we should prove that S is a linearly independent set of vectors.

Now we should show that S spans F [x] *i.e.*, each polynomial in F [x] can be expressed as a linear combination of a finite number of elements of S.

Let $f(x) = a_0 + a_1x + a_2x^2 + ... + a_tx^t$ be a polynomial of degree t.

Then $f(x) = (a_0)1 + a_1x + a_2x^2 + ... a_tx_t$

Hence S is a basis of F [x].

Note: The vector space F [x] has no finite basis. If we take any finite set W of polynomials, we can find a polynomial of degree greater than that of each of them. Such a polynomial cannot at any cost be expressed as a linear combination of the elements of W.

2.11 FINITE DIMENSIONAL VECTOR SPACE

Definition: *The vector space V(F) is said to be* **finite dimensional** *or* **finitely generated** *if there exists a* **finite** *subset of S of V such that V = L (S).*

The vector space V_n (F) of n-tuples is a finite dimensional vector space.

The vector space F[x] of all polynomials over a field F is not finite dimensional. There exists no finite subset S of F [x] which spans F[x]. *A vector space which is no finitely generated may be referred to as an infinite dimensional space.* Thus, the vector space F[x] of all polynomials over a field F is infinite dimensional.

Existence of Basis of a Finite Dimensional Vector Space

Theorem

There exists a basis for each finite dimensional vector space.

Proof:

Let V(F) be a finitely generated vector space. Let S $\{\alpha_1, \alpha_2, ..., \alpha_m\}$ be a finite subset of V such that L(S) = V. Without loss of generality we may suppose that no member of S is 0.

If S is linearly independent, then S itself is a basis of V. If S is linearly dependent, then there exists, a vector $\alpha_t \in S$ which can be expressed as a linear combination of the preceding vectors $\alpha_1, \alpha_2, ..., \alpha_{i-1}$.

If we omit this vector α_i from S, then the remaining set S' of m–1 vectors

$$\alpha_1, \alpha_2, ..., \alpha_{i-1}, \alpha_{i+1}, ..., \alpha_m$$

also generates V *i.e.* V = L(S'). For if α is any element of V, then L(S) = V implies that α can be written as a linear combination of $\alpha_1, \alpha_2, ...,$

α_m. In this linear combination we can replace α_i by a linear combination of $\alpha_1, \alpha_2, ..., \alpha_{i-1}$. Then α will be a linear combination of $\alpha_1, \alpha_2, ..., \alpha_{i-1}, \alpha_{i+1} ...,$ α_m. Thus V will be equal to L(S').

If S' is linearly independent, then S' will be a basis of V. If S' is linearly dependent, then proceeding as above are shall get a now set of m–2 vectors which generates V. Continuing this process we shall, after a finite number of steps, obtain a linearly independent subset of S which generates V and which is therefore a basis of V.

At the most it may happen that we shall be left with a subset of S which contains only one non-zero vector and which spans V. We know that a set containing a single non-zero vector is definitely linearly independent and so it will form a basis of V.

Remark: *The above theorem may also be stated as below:*

If a finite set S of vectors spans a finite dimensional vector space V(F), there exists a subset of S which forms a basis of V.

Invariance of Number of Elements in the Basis of a Finite Dimensional Vector Space

Theorem:

If V(F) is a finite dimensional vector space, then any two bases of V have the same number of elements.

Proof:

Suppose V(F) is a finite dimensional vector space. Then V definitely possesses a basis. Let $S_1 = \{\alpha_1, \alpha_2, ..., \alpha_m\}$ and $S_2 = \{\beta_1, \beta_2, ..., \beta_n\}$ be two bases of V. We shall prove that m = n.

Since $V = L(S_1)$ and $\beta \in V$, therefore β_1 can be expressed as a linear combination of $\alpha_1, \alpha_2, ..., \alpha_m$. Consequently the set $S_3 = \{\beta_1, \alpha_1, \alpha_2, ..., \alpha_m\}$ which also obviously generates V(F) is linearly dependent. Therefore there exists a member $\alpha_i \neq \beta_1$ of this set S_3 such that α_i is a linear combination of the preceding vectors $\beta_1, \alpha_1, \alpha_2, ..., \alpha_{i-1}$. If we omit the vector α_i from S_3 then V is also generated by the remaining set

$$S_4 = \{\beta_1, \alpha_1, \alpha_2, ..., \alpha_{i-1}, \alpha_{i+1}, ..., \alpha_m\}.$$

Since $V = L(S_4)$ and $\beta_2 \in V$, therefore β_2 can be expressed as a linear combination of the vectors belonging to S_4. Consequently the set

$$S_5 = \{\beta_2, \beta_1, \alpha_1, \alpha_2, ..., \alpha_{i-1}, \alpha_{i+1}, ..., \alpha_m\}$$

is linearly dependent. Therefore there exists a member α_j of this set S_5 such that α_j is a linear combination of the preceding vectors. Obviously α_j will be different from β_1 and β_2 since $\{\beta_1, \beta_2\}$ is a linearly independent set. If we exclude the vector α_j from S_5, then the remaining set will generate V(F).

We may continue to proceed in this manner. Here each step consists in the exclusion of an α and the inclusion of a β in the set S_1.

Obviously the set S_1 of α's cannot be exhausted before the set S_2 of β's, otherwise V(F) will be a linear span of a proper subset of S_2 and thus S_2 will become linearly dependent. Therefore we must have $m \notin n$

Interchanging the roles of S_1 and S_2 we shall get that

$$n < m. \text{ Hence } m = n.$$

Example:

For the vector space $V_3(F)$, both the sets

$$S_1 = \{(1, 0, 0), (0, 1, 0), (0, 0, 1)\}$$

and $$S_2 = \{(1, 0, 0), (1, 1, 0), (1, 1, 1)\}$$

are bases as can be easily seen. Both these contain the same number of elements *i.e.*, 3.

Dimension of a Finitely Generated Vector Space

Definition. *The number of elements in any basis of a finite dimensional vector space V(F) is called the dimension of the vector space V(F) and will be denoted by dim V.*

The vector space $V_n(F)$ is of dimension n. The vector space $V_3(F)$ is of dimension 3. If a field F is regarded as a vector space over F, then F will be of dimension 1 and the see S = (1) consisting of unity element of F alone is a basis of F. In fact every non-zero element of F will form a basis of F.

2.12 SOME PROPERTIES OF FINITE DIMENSIONAL VECTOR SPACES

Theorem 1: *(Extension Theorem)*

Every linearly independent subset of a finitely generated vector space V(F) forms a part of a basis of V.

Or

Every linearly independent of a finitely generated vector space V(F) is either a basis of V or can be extended to form a basis of V.

Proof:

Let $S = \{\alpha_1, \alpha_2, ..., \alpha_m\}$ be a linearly independent subset of a finite dimensional vector space V(F). If dim V = n, then V has a finite basis say, $\{\beta_1, \beta_2, ..., \beta_n\}$. Consider the set

$$S_1 = \{\alpha_1, \alpha_2, ..., \alpha_m, \beta_1, \beta_2, ..., \beta_n\}.$$

Obviously $L(S_1) = V$. Since the α's can be expressed as linear combination of the β's therefore the set S_1 is linearly dependent.

Therefore there is some vector of S_1 which is a linear combination of its preceding vectors. This vector cannot be any of the α's since the α's are linearly independent. Therefore this vector must be some β, say, β_i. Now omit the vector β_i from S_i and consider the set

$$S_2 = \{\alpha_1, \alpha_2, ..., \alpha_m, \beta_1, \beta_2, ..., \beta_{i-1}, \beta_{i+1}, ..., \beta_n\}.$$

Obviously $L(S_2) = V$. If S_2 is linearly independent then S_2 will be a basis of V and it is the required extended set which is a basis of V. If S_2 is not linearly independent, then repeating the above process a finite number of times, we shall get a linearly independent set containing $\alpha_1, \alpha_2, ..., \alpha_m$ and spanning V. This set will be basis of V and it will contain S. Since each basis of V contains the same number of elements, therefore exactly n – m elements of the set of β's will be adjoined to S so as to form a basis of V.

Theorem 2:

Let $S = \{\alpha_1, \alpha_2, ..., \alpha_n\}$ be a basis of a finite dimensional vector space V(F) of dimension n. Then every element α of V can be uniquely expressed as $\alpha = a_1\alpha_1 + a_2\alpha_2 + ... + a_n\alpha_n$ where $a_1, a_2, ..., a_n \in F$.

Proof:

Since S is a basis of V, therefore L(S) = V. Therefore any vector $\alpha \in V$ can be expressed as $\alpha = a_1\alpha_1 + a_2\alpha_2 + ...1 + a_n\alpha_n$.

The show uniqueness, let us suppose that

$$\alpha = b_1\alpha_1 + b_2\alpha_2 + ... + b_n\alpha_n.$$

Then we must show that $a_1 = b_1, a_2 = b_2, ..., a_n = b_n$

We have $a_1\alpha_1 + a_2\alpha_2 + ... + a_n\alpha_n = b_1\alpha_1 + b_2\alpha_2 + ... + b_n\alpha_n$

$\Rightarrow \quad (a_1 - b_1)\alpha_1 + (a_2 - b_2)\alpha_2 + ... + (a_n - b_n)\alpha_n = 0$

$\Rightarrow \quad a_1 - b_1 = 0, a_2 - b_2 = 0, ..., a_n - b_n = 0$ since $\alpha_1, \alpha_2, ..., \alpha_n$ are linearly independent

$\Rightarrow \quad a_1 = b_1, a_2 = b_2, ..., a_n = b_n.$

Hence the theorem.

Theorem 3:

If W_1, W_2 are two subspaces of a finite dimensional vector space V(F), then

$$dim\ (W_1 + W_2) = dim\ W_1 + dim\ W_2 + dim\ (W_1 \cap W_2).$$

Proof:

Let dim $(W_1 \cap W_2) = k$ and let the set

$$S = \{\gamma_1, \gamma_2, \gamma_3, ..., \gamma k\}$$

be a basis of $W_1 \cap W_2$. The $S \subseteq W_1$ and $S \subseteq W_2$.

Since S is linearly independent and $S \subseteq W_1$, therefore S can be extended to form a basis of W_1. Let

$$\{\gamma_1, \gamma_2, ..., \gamma_k, \alpha_1, \alpha_2, ..., \alpha_m\}$$

be a basis of W_1. Then dim $W_1 = k + m$. Similarly let

$$\{\gamma_1, \gamma_2, ..., \gamma_k, \beta_1, \beta_2, ..., \beta_t\}$$

be a basis of W_2. Then dim $W_2 = k + t$.

$$\therefore \dim W_1 + \dim W_2 - \dim (W_1 \cap W_2) = (m + k) + (k + t) - k$$
$$= k + m + t.$$

$\therefore$ to prove the theorem we must show that

$$\dim (W_1 + W_2) = k + m + t.$$

We claim that the set

$$S_1 = \{\gamma_1, \gamma_2, ..., \gamma k, \alpha_1, \alpha_2, ..., \alpha_m, \beta_1, \beta_2, ..., \beta_1\}$$

is a basis of $W_1 + W_2$.

First we show that S_1 is linearly independent. Let

$$c_1\gamma_1 + c_2\gamma_2 + ... + c_k\gamma_k + a_1\alpha_1 + a_2\alpha_2 + ... + a_m\alpha_m + b_1\beta_1 + b_2\beta_2 + ... + b_t\beta_t = 0 \quad ...(1)$$

$$\Rightarrow b_1\beta_1 + b_2\beta_2 + ... + b_t\beta_t = -(c_1\gamma_1 + ... + c_k\gamma_k + a_1\alpha_1 + ... + a_m\alpha_m) \quad ...(2)$$

Now $-(c_1\gamma_1 + ... + c_k\gamma_k + a_1\alpha_1 + ... + a_m\alpha_m) \in W_1$ since it is a linear combination of a basis of W_1. Again

$$b_1\beta_1 + b_2\beta_2 + ... + b_t\beta_t \in W_2$$

since it is a linear combination of elements belonging to a basis of W_2.

Also by virtue of the equality (2), $b_1\beta_1 + ... + b_t\beta_t \in W_1$. Therefore $b_1\beta_1 + b_2\beta_2 + ... + b_t\beta_t \in W_1 \cap W_2$. Therefore it can be expressed as a linear combination of the basis of $W_1 \cap W_2$. Thus we have a relation of the form

$$b_1\beta_1 + b_2\beta_2 + ... + b_t\beta_t = d_1\gamma_1 + d_2\gamma_2 + ... + d_k\gamma_k$$

$$\Rightarrow \quad b_1\beta_1 + b_2\beta_2 + ... + b_t\beta_t - d_1\gamma_1 - d_2\gamma_2 - ... - d_k\gamma_k = 0.$$

But $\beta_1, \beta_2, ..., \beta_t, \gamma_1, ..., \gamma_k$ are linearly independent vectors.

Therefore we must have $b_1 = 0, b_2 = 0, \ldots, b_t = 0$.

Putting these values of β's in (1), it reduces to

$$c_1\gamma_1 + c_2\gamma_2 + \ldots + c_k\gamma_k + a_1\alpha_1 + a_2\alpha_2 + \ldots + a_m\alpha_m = 0$$

$$\Rightarrow \quad c_1 = 0, c_2 = 0, \ldots, c_k = 0, a_1 = 0, a_2 = 0, \ldots, a_m = 0$$

since the vectors $\gamma_1, \gamma_2, \ldots, \gamma_k, \alpha_1, \alpha_2, \ldots, \alpha_m$ are linearly independent.

Thus the relation (1) implies that

$$c_1 = 0, c_2 = 0, \ldots, c_k = 0, a_1 = 0, \ldots, a_m = 0, b_1 = 0, \ldots, b_t = 0.$$

Therefore the set S_1 of vectors $\gamma_1, \ldots, \gamma_k, \alpha_1, \ldots, \alpha_m, \beta_1, \ldots, \beta_t$ is linearly independent.

Now to show that $L(S_1) = W_1 + W_2$.

Since $W_1 + W_2$ is a subspace of V and each element of S_1 belongs to $W_1 + W_2$, therefore $L(S_1) \subseteq W_1 + W_2$.

Again let α be any element of $W_1 + W_{-2}$. Then

α = some element of W_1 + some element of W_2

= a linear combination of a basis of W_1 + a linear combination of a basis of W_2

= a linear combination of S_1.

$\therefore \alpha \in L(S_1)$. Hence $W_1 + W_2 \subseteq L(S_1)$.

$\therefore L(S_1) = W_1 + W_2$.

$\therefore S_1$ is a basis of $W_1 + W_2$ and consequently

$\dim(W_1 + W_2) = k + m + t$.

Hence the theorem.

Dimension of a Subspace

Theorem 1:

Each subspace W of a finite dimensional vector space V(F) of dimension n is a finite dimensional space with dim $m \leq n$.

Also V = W iff dim V = dim W.

Proof:

Let V(F) be a finite dimensional vector space of dim n. Let W be a subspace of V. Any subset of W containing (n + 1) or more vectors is also a subset of V and any (n + 1) vectors in V are linearly dependent. Therefore any linearly independent set of vectors in W can contain, at the most n vectors. Let

$$S = \{\alpha_1, \alpha_2, ..., \alpha_m\}$$

be a linearly independent subset of W with a maximal number of elements. We claim that S is a basis of W. The proof is as follows:

(i) By assumption S is a linearly independent subset of W.

(ii) L (S) = W as we shall just show.

Let α be any element of W.

Then the (m + 1) vectors $\alpha, \alpha_1, ..., \alpha_m$ belonging to W are linearly dependent because we have supposed that the largest independent subset of W contains m vectors.

Now $\{\alpha_1, \alpha_2, ..., \alpha_m, \alpha\}$ is a linearly dependent set. Therefore there exists a vector belonging to it which can be expressed as a linear combination of the preceding vectors. Since $\alpha_1, \alpha_2, ..., \alpha_m$ are linearly independent, therefore this vector cannot be any of these m vectors. So it must be α itself. Thus, α can be expressed as a linear combination of $\alpha_1, \alpha_2, ..., \alpha_m$.

Hence L(S) = W.

$\therefore$ S is a basis of W.

$\therefore$ dim W = m and $m \leq n$.

Now if V = W, then every basis of V is also a basis of W.

Hence dim V = dim W = n.

Conversely let dim W = dim V = n. Then to prove that W = V.

Let S be a basis of W. Then L(S) = W and S contains n vectors. Since S is also a subset of V and S contains n linearly independent vectors, therefore S will also be a basis of V. Therefore L(S) = V. Hence W = V.

Theorem 2:

Each set of (n + 1) or more vectors of a finite dimensional vector space V(F) of dimension n is linearly dependent.

Proof:

Let V(F) be a finite dimensional vector space of dimension n. Let S be a linearly independent subset of V containing (n + 1) or more vectors. Then S can be extended to form a basis of V. Thus we shall bet a basis of V containing more than n vectors. But every basis of V will contain exactly n vectors. Hence, our assumption is wrong. Therefore if S contains (n + 1) or more vectors, then S must be linearly dependent. From this theorem we conclude that if S contains m vectors and S is linearly independent then

$$m \leq n.$$

Theorem 3:

If V(F) is a finite dimensional vector space of dimension n, then any set of a n linearly independent vectors in V forms a basis of V.

Proof:

Let S = $\{\alpha_1, \alpha_2, ..., \alpha_n\}$ be a linearly independent subset of finite dimensional vector space V(F) of dimension n. If S is not a basis of V, then it can be extended to form a basis of V. Thus we shall get a basis of V containing more than n vectors. But every basis of V must contain exactly n vectors. Therefore our assumption is wrong and S must be a basis of V.

Theorem 4:

If a set S of n vectors of a finite dimensional vector space V(F) of dimension n generators V(F), then S is a basis of V.

Proof:

Let V(F) be a finite dimensional vector space of dimension n. Let S = $(\alpha_1, \alpha_2, ..., \alpha_n\}$ be a subset of V such that L(S) = V.

If S is linearly independent, then S will form a basis of V. If S is not linearly independent, then there will exist a proper subset of S which will form a basis of V. Thus we shall get a basis of V containing less than n elements. But every basis of V must contain exactly n elements. Hence S cannot be linearly dependent and so S must be a basis of V.

Note: *If is a finite dimensional vector space of dimension n, then V cannot be generated by fewer than n vectors.*

Example 1:

Show that the vectors (1, 2, 1), (2, 1, 0), (1, – 1, 2) form a basis of $\mathbf{R}^3$.

Solution:

We know that the set {(1, 0, 0), (0, 1, 0), (0, 0, 1) forms a basis for $\mathbf{R}^3$. Therefore dim $\mathbf{R}^3 = 3$. If we show that the set S = {(1, 2, 1,), (2, 1, 0), (1, – 1, 2)} is linearly independent, then this will also form a basis for $\mathbf{R}^3$.

We have

$$a_1 (1, 2, 1) + a_2 (2, 1, 0) + a_2 (1, -1, 2) = (0, 0, 0)$$

$$\Rightarrow \quad (a_1 + 2a_2 + a_3, 2a_1 + a_2 - a_3, a_1 + 2a_3) = (0, 0, 0).$$

$$\therefore \quad a_1 + 2a_2 + a_3 = 0 \qquad ...(1)$$

$$2a_1 + a_2 - a_3 = 0 \qquad ...(2)$$

$$a_1 + 2a_3 = 0 \qquad ...(3)$$

Now, we shall solve these equations to get the values of a_1, a_2, a_3. Multiplying the equation (2) by 2, we get

$$4a_1 + 2a_2 - 2a_3 = 0. \qquad ...(4)$$

Subtracting (4) from (1), we get

$$-3a_1 + 3a_3 = 0.$$

$$\Rightarrow \qquad -a_1 + a_3 = 0. \qquad ...(5)$$

Adding (3) and (5), we get $3a_3 = 0 \Rightarrow a_3 = 0$. Putting $a_3 = 0$ in (3), we get $a_1 = 0$. Now putting $a_3 = 0$ and $a_1 = 0$ in (1), we get $a_2 = 0$.

Thus, solving the equations (1), (2) and (3), we get $a_1 = 0$, $a_2 = 0$, $a_3 = 0$. Therefore the set S is linearly independent. Hence it forms a basis of $\mathbf{R}^3$.

Example 2:

Select a basis, if any, of $\mathbf{R}^3(\mathbf{R})$ from the set

$\{\alpha_1, \alpha_2, \alpha_3, \alpha_4\}$, where $\alpha_1 = (1, -3, 2)$, $\alpha_2 = (2, 4, 1)$, $\alpha_3 = (3, 1, 3)$, $\alpha_4 = (1, 1, 1)$.

Solution:

Let $S = \{\alpha_1, \alpha_2, \alpha_3, \alpha_4\}$.

If any three vectors in S are linearly independent, then they will form a basis of the vector space $\mathbf{R}^3$ $(\mathbf{R})$,

First consider the set $S_1 = \{\alpha_1, \alpha_2, \alpha_3\}$. Let us see whether the vectors in the set S_1 are linearly independent or not

The determinant of order 3 whose columns consist of the coordinates of the vectors α_1, α_2, α_3 is

$$= \begin{vmatrix} 1 & 2 & 3 \\ -3 & 4 & 1 \\ 2 & 1 & 3 \end{vmatrix}$$

$$= \begin{vmatrix} 1 & 0 & 0 \\ -3 & 10 & 10 \\ 2 & -3 & -3 \end{vmatrix}, \text{ by } C_2 - 2C_1 \text{ and } C_3 - 3C_1$$

$$= -30 + 30 = 0.$$

$\therefore$ the vectors $\alpha_1, \alpha_2, \alpha_3$ are linearly dependent and so they do not form a basis of $\mathbf{R}^3(\mathbf{R})$.

Now consider the set $S_2 = \{\alpha_1, \alpha_2, \alpha_4\}$.

The determinant of order 3 whose columns consist of the coordinates of the vectors $\alpha_1, \alpha_2, \alpha_4$ is

$$= \begin{vmatrix} 1 & 2 & 1 \\ -3 & 4 & 1 \\ 2 & 1 & 1 \end{vmatrix}$$

$$= \begin{vmatrix} 1 & 0 & 0 \\ -3 & 10 & 4 \\ 2 & -3 & -1 \end{vmatrix}, \text{ by } C_2 - 2C_1 \text{ and } C_2 - C_1$$

$= -10 + 12 = 2$ *i.e.*, $\neq 0$.

$\therefore$ the vectors $\alpha_1, \alpha_2, \alpha_4$ are linearly independent.

Since $S_2 = \{\alpha_1, \alpha_2, \alpha_4\}$ is a linearly independent subset of $\mathbf{R}^3$ containing three vectors, therefore it is a basis of $\mathbf{R}^3$.

Example 3:

Show that the set $S = \{1, x, x^2,..., x^n\}$ of $n + 1$ polynomials in x is a basis of the vector space $P_n(\mathbf{R})$, of all polynomials in x (of degree at most n) over the field of real numbers.

Solution:

$P^n(\mathbf{R})$ is the vector space of all polynomials in x (of degree at most n) over the field $\mathbf{R}$ of real numbers.

$S = \{1, x, x^2,..., x^n)$ is a subset of P_n consisting of n + 1 polynomials. To prove that S is a basis of the vector space $P_n(\mathbf{R})$.

First we show that the vectors in the set S are linearly independent over the field $\mathbf{R}$.

The zero vector of the vector space $P_n(\mathbf{R})$ is the zero polynomial. Let $a_0, a_1, a_2,..., a_n \in R$ be such that

$a_0 (1) + a_1x + a_2x^3 + ... + a_nx^n = 0$ *i.e.*, zero polynomial.

Now by the definition of the equality of two polynomials, we have

$$a_0 + a_1x + a_2x^2 + ... a_nx^n = 0$$

$$\Rightarrow \quad a_0 = 0, a_1 = 0, a_2 = 0,..., a_n = 0.$$

$\therefore$ the vectors 1, x, x^2,..., x^n of the vector space $P_n(\mathbf{R})$ are linearly independent.

Now we shall show that the set S generates the whole vector space $P_n(\mathbf{R})$.

Let $\alpha = a_0 = a_1x + a_2x^2 + \ldots a_nx^n$ be any arbitrary member of P_n, where $a_0, a_1, \ldots, a_n \in \mathbf{R}$. The α is a linear combination of the polynomials 1, x, x^2,..., x^n over the field **R**. Therefore S generates $P_n(\mathbf{R})$.

Since S is a linearly independent subset of $P_n(\mathbf{R})$ and it also generates $P_n(\mathbf{R})$, therefore it is a basis of $P_n(\mathbf{R})$.

Example 4:

Prove that any finite set S of vectors, not all the zero vectors, contains a linearly independent subset T which spans the same space as S.

Solution:

Let S be a finite set of vectors belonging to a vector space V(F). Assume that no member of S is zero vector for if any member of S is zero vector, we can omit it from S without affecting the subspace spanned by S.

Let $S = \{\alpha_1, \alpha_2, \ldots, \alpha_m\}$.

If S is linearly independent, then S itself is the required linearly independent subset T of S which spans the same subspace of V as S.

If S is linearly dependent, then there exists, a vector $\alpha_1 \in S$ which can be expressed as a linear combination of the preceding vectors $\alpha_1, \alpha_2 \ldots, \alpha_{i-1}$.

If we omit this vector α_i from S, then the remaining subset S' of S containing m – 1 vectors spans the same subspace of V as S.

If S' is linearly independent, then S' will be the required linearly independent subset of S which spans the same subspace of V as S. If S' is linearly dependent, then proceeding as above we shall get a new subset of S. Continuing this process we shall, after a finite number of steps, obtain a linearly independent subset of S which spans the same space as S.

At the most it may happen that we shall be left with a subset of S which contains only one non-zero vector and which spans the same space as S. We know that a set containing a single non-zero vector is definitely linearly independent.

Hence any finite set S of vectors, not all the zero vectors, definitely contains a linearly independent subset T which spans the same space as S.

Example 5:

Let V be a vector space. Let W be a subspace of V generated by the vectors $\alpha_1,\ldots, \alpha_3$. Prove that W is spanned by a linearly independent subset of $\alpha_1,\ldots, \alpha_3$.

Solution:

W is a subspace of V generated by a finite set

$$S = \{\alpha_1,\ldots, \alpha_2\}.$$

To show that there exists a linearly independent subset T of S which also spans W.

Example 6:

If W is a subspace of a finite dimensional vector space V, prove that any basis of W can be extended to form a basis of V.

Solution:

Let V (F) be a finite dimensional vector space of dimension n. Let $\{\beta_1, \beta_2,\ldots, \beta_n\}$ be a basis of V.

Let W be subspace of V. Then W itself is finite dimensional and dim W $\leq$ n. Let dim W = m and let $S = \{\alpha_1, \alpha_2,\ldots, \alpha_m\}$ be a basis of W.

Then S is a linearly independent subset of V. To show that S can be extended to form a basis of V.

Example 7:

If n vectors span a vector space V containing r linearly independent vectors, then show that $n \geq r$.

Solution:

Suppose a subset S of V containing n vectors spans V. Then there exists a linearly independent subset T of S which also spans V. This subset T of S will form a basis of V. Suppose the number of vectors in T is *m*. Then dim V = m $\leq$ n.

Since dim V = m, therefore any subset of V containing more than m vectors will be linearly dependent.

Hence if S_1 is a linearly independent subset of V and S_1 contains r vectors, we must have

$$r \leq m \quad \Rightarrow \quad r \leq n. \qquad [\because m \leq n]$$

Example 8:

Show that a finite subset W of a vector space V(F) is linearly dependent if and only if some element of W can be expressed as a linear combination of the others.

Solution:

Let $W = \{\alpha_1, \alpha_2, ..., \alpha_n\}$ be a finite subset of a vector space V(F).

First suppose that W is linearly dependent. Then to show that some vector of W can be expressed as a linear combination of the others.

Since the vectors $\alpha_1, \alpha_2, ..., \alpha_n$ are linearly dependent, therefore there exist scalars $a_1, a_2, ..., a_n$ not all zero such that

$$a_1\alpha_1 + a_2\alpha_2 + ... + a_n\alpha_n = 0. \qquad ...(1)$$

Suppose $a_r \neq 0$, $1 \leq r \leq n$.

Then from (1), we have

$$a_r\alpha_r = -a_1\alpha_1 - a_2\alpha_2 - ... - a_{r-1}\alpha_{r-1} - a_{r+1}\alpha_{r+1} - ... - a_n\alpha_n$$

$$\Rightarrow \quad a_r^{-1}(a_r\alpha_r) = a_r^{-1}(-a_1\alpha_1 - a_2\alpha_2 - ... - a_{r-1}\alpha_{r-1} - a_{r+1}\alpha_{r+1} - ... - a_n\alpha_n)$$

$$\Rightarrow \quad \alpha_r = (-a_r^{-1}a_1)\alpha_1 + (-a_r^{-1}a_2)\alpha_2 + ... + (-a_r^{-1}a_{r-1})\alpha_{r-1}$$

$$+ (a_r^{-1}a_{r+1})\alpha_{r+1} + ... + (-a_r^{-1}a_n)\alpha_n$$

Thus, α_r is a linear combination of the other vectors of the set W.

Conversely suppose that some vector of W is a linear combination of the others. Then to prove that W is linearly dependent.

Without loss of generality suppose α_1 is a linear combination of the vectors $\alpha_2, ..., \alpha_n$.

Let $\alpha_1 = b_2\alpha_2 + b_3\alpha_3 + ... + b_n\alpha_n$, where $b_2, b_2, ..., b_n \in F$.

Then $\quad \alpha_1 - b_2\alpha_2 - b_3\alpha_3 - ... - b_n\alpha_n = 0. \qquad ...(2)$

Since in the linear relation (2) among the vectors $\alpha_1, \alpha_2, ..., \alpha_n$, the scalar coefficient of α_1 is 1 which is $\neq 0$, therefore the vectors $\alpha_1, \alpha_2, ..., \alpha_n$ are linearly dependent.

Example 9:

Determine whether or not the following vectors form a basis of $\mathbf{R}^3$*:*

(1, 1, 2), (1, 2, 5), (5, 3, 4).

Solution:

We know that dim $\mathbf{R}^3 = 3$. If the given set of vectors is linearly independent, it will form a basis of $\mathbf{R}^3$ otherwise not. We have

$$a_1 (1, 1, 2) + a_2 (1, 2, 5) + a_2 (5, 3, 4) = (0, 0, 0)$$

$$\Rightarrow (a_1 + a_2 + 5a_3, a_1 + 2a_2 + 3a_3, 2a_1 + 5a_2 + 4a_2) = (0, 0, 0).$$

$\therefore$ $$a_1 + a_2 + 5a_3 = 0 \quad ...(1)$$

$$a_1 + 2a_2 + 3a_3 = 0 \quad ...(2)$$

$$2a_1 + 5a_2 + 4a_3 = 0. \quad ...(3)$$

Now, we shall solve these equations to get the values of a_1, a_2, a_3. Subtracting (2) from (1), we get

$$- a_2 + 2a_3 = 0. \quad ...(4)$$

Multiplying (1) by 2, we get

$$2a_1 + 2a_2 + 10a_2 = 0. \quad ...(5)$$

Subtracting (5) from (3), we get

$$3a_2 - 6a_3 = 0$$

$\Rightarrow$ $$a_2 - 2a_3 = 0. \quad ...(6)$$

We see that the equations (4) and (6) are the same and give $a_2 = 2a_3$. Putting $a_2 = 2a_3$ in (1), we get $a_1 = - 7a_3$. If we put $a_2 = 1$ we get $a_2 = 2$ and $a_1 = - 7$. Thus, $a_1 = - 7$, $a_2 = 2$, $a_3 = 1$ is a non-zero solution of the equations (1), (2) and (3). Hence the give set is linearly dependent so it does not form a basis of $\mathbf{R}^2$.

Example 10:

Show that the set {(1, i, 0), (2i, 1, 1), (0, 1 + i, 1 – i)} is a basis for $V_3(C)$.

Solution:

We know that dim $V_3(C)$ or $C^3 = 3$. If the given set containing three vectors is linearly independent, it will form a basis of V_3 (C) otherwise not.

Let a, b, c $\in$ C be such that

$$a (1, i, 0) + b (2i, 1, 1) + c (0, 1, + i, 1 - i) = (0, 0, 0)$$

$$\Rightarrow (a + 2ib + 0c, ai + b + c [1 + i], 0a + b + c [1 \quad i]) - (0, 0, 0).$$

$\therefore$ $$a + 2ib = 0, \quad ...(1)$$

$$ai + b + c (1 + i+ = 0, \quad ...(2)$$

and $$b + c (1 - i) = 0. \quad ...(3)$$

Now we shall solve these equations to get the values of a, b, c.

Multiplying (1) y – i and adding to (2), we get

$$3b + c (1 + i) = 0. \quad ...(4)$$

Multiplying (3) by 3 and subtracting from (4), we get

$$c\,(1 + i) - 3c\,(1 - i) = 0$$

$$\Rightarrow \quad c\,(1 + i - 3 + 3i) = 0 \Rightarrow c\,(-2 + 4i) = 0 \Rightarrow c = 0.$$

Putting c = 0 in (3), we get b = 0

Putting b = 0 in (1), we get a = 0.

Thus, the only solution of the equations (1), (2) and (3) is a = 0, b = 0, c = 0. Therefore the three given vectors are linearly independent and so they form a basis of $V_3(C)$.

Example 11:

For the 3-dimensional space $\mathbf{R}^3$ *over the field of real numbers* **R**, *determine if the set {(2, – 1, 0), (3, 5, 1), (1, 1, 2)} is a basis.*

Solution:

We have dim $\mathbf{R}^3 = 3$. If the given set containing three vectors is linearly independent, it will form a basis of $\mathbf{R}^3$ otherwise not.

Let a, b, c $\in$ **R** be such that

$$a\,(2, -1, 0) + b\,(3, 5, 1) + c\,(1, 1, 2) = (0, 0, 0)$$

$$\Rightarrow \quad (2a + 3b + c, -a + 5b + c, 0a + b + 2c) = (0, 0, 0).$$

$$\therefore \quad 2a + 3b + c = 0, \quad \text{...(1)}$$

$$-a + 5b + c = 0, \quad \text{...(2)}$$

$$\text{and} \quad b + 2c = 0. \quad \text{...(3)}$$

Now we shall solve these equations to get the values of a, b, c.

Multiplying (2) by 2 and adding to (1), we get

$$13b + 3c = 0. \quad \text{...(4)}$$

Multiplying (3) by 13 and then subtracting (4) from it, we get

$$23c = 0 \Rightarrow c = 0.$$

Putting c = 0 in (3), we get b = 0.

Putting b = 0, c = 0 in (1), we get a = 0.

Thus, the only solution of the equations (1), (2) and (3) is a = 0, b = 0, c = 0. Therefore the three given vectors are linearly independent and so they form a basis of $\mathbf{R}^3$.

Example 12:

Show that a system X consisting of the vectors

$\alpha_1 = (1, 0, 0, 0)$, $\alpha_2 = (0, 1, 0, 0)$, $\alpha_3 = (0, 0, 1, 0)$ and $\alpha_4 = (0, 0, 0, 1)$ is a basis set of **R^4 (R).**

Solution:

First we show that the set X is a linearly independent set of vectors.

If a_1, a_2, a_3, a_4 be any scalars *i.e.*, elements of the field R, then

$$a_1\alpha_1 + a_2\alpha_2 + a_3\alpha_3 + a_4\alpha_4 = \text{zero vector}$$

$$\Rightarrow a_1 (1, 0, 0, 0) + a_2 (0, 1, 0, 0) + a_3 (0, 0, 1, 0) + a_4 (0, 0, 0, 1) = (0, 0, 0, 0)$$

$$\Rightarrow (a_1, a_2, a_3, a_4) = (0, 0, 0, 0)$$

$$\Rightarrow a_1 = 0, a_2 = 0, a_3 = 0, a_4 = 0.$$

Therefore the given set X of four vectors is linearly independent.

Now we shall that X generates $\mathbf{R}^4$ *i.e.*, each vector of $\mathbf{R}^4$ can be expressed as a linear combination of the vectors of X.

Let (a, b, c, d) be any vector in $\mathbf{R}^4$. We can write

$$(a, b, c, d) = a(1, 0, 0, 0) + b(0, 1, 0, 0) + c(0, 0, 1, 0) + d(0, 0, 0, 1)$$
$$= a\alpha_1 + b\alpha_2 + c\alpha_3 + d\alpha_4.$$

Thus (a, b, c, d) has been expressed as a linear combination of the vectors of X and so X generates $\mathbf{R}^4$.

Since X is a linearly independent subset of $\mathbf{R}^4$ and it also generates R^4, therefore it is a basis of $\mathbf{R}^4$.

2.13 HOMOMORPHISM OF VECTOR SPACES OR LINEAR TRANSFORMATIONS

Definition: *Let U(F) and V(F) be two vector spaces. Then a mapping $f: U \to V$ is called a homomorphism or a linear transformation of 'U into V if:*

$$\text{(i) } f(\alpha + \beta) = f(\alpha) + f(\beta), \ \forall\ \alpha, \beta, \in U$$

$$\text{and (ii) } f(a\alpha) = af(\alpha)\ \forall\ a \in F,\ \forall\ \alpha \in U.$$

The conditions (i) and (ii) can be combined into a single condition $f(a\alpha + b\beta) = af(\alpha) + bf(\beta)\ \forall\ a, b \in F$ and $\forall\ \alpha, \beta, \in U$.

If F is a homomorphism of U onto V, then V is called a homomorphic image of U.

Theorem 1:

If f is a homomorphism of U(F) into V(F), then

(i) f(0) = 0' where 0 and 0' are the zero vectors of U and V respectively.

(ii) f(– α) = – f(α) ∀ α ∈ U.

Proof:

(i) Let $\alpha \in U$. Then $f(\alpha) \in V$. Since 0' is the zero vector of V, therefore $f(\alpha) + 0' = f(\alpha) + f(\alpha + 0) = f(\alpha) + f(\alpha) + f(0)$

Now V is an abelian group with respect to addition of vectors. Therefore $f(\alpha) + 0' = f(\alpha) + f(0) \Rightarrow 0' = f(0)$ by left cancellation law.

(ii) If $\alpha \in U$, then $-\alpha \in U$. Also we have

$$0' = f(0) = f[\alpha + (-\alpha)] = f(\alpha) + f(-\alpha).$$

Now $f(\alpha) + f(-\alpha) = 0' \Rightarrow f(-\alpha) =$ additive inverse of $f(\alpha)$

$$\Rightarrow f(-\alpha) = -f(\alpha).$$

Remark: The students may use without any confusion the same symbol 0 to denote both the zero vectors of U and V. In the relation $f(0) = 0$, the zero in the left hand side is the zero vector of U and its f-image *i.e.*, the zero on the right hand side is the zero vector of V.

Theorem 2:

The kernel of a homomorphism is a subspace.

Proof:

Let U(F) and V(F) be two vector spaces and let f be a homomorphism of U into V.

Let 0' be the zero vector of V and 0 be the zero vector of U. Let W be the kernal of f *i.e.* $W = \{\alpha \in U: f(\alpha) = 0'\}$.

To prove that W is a sub space of U.

Let α, β be any two elements of W. Then $f(\alpha) = 0'$ and $f(\beta) = 0'$.

If α, β are any two elements of F, then we have

$f(a\alpha + b\beta) = af(\alpha) + bf(\beta)$ [$\because$ f is a homomorphism of U into V]

$= a0' + b0' = 0' + 0' = 0'$.

$\therefore a\alpha + b\beta \in W$.

Thus $a, b \in F$ and $\alpha, \beta, \in W \Rightarrow a\alpha + b\beta \in W$.

$\therefore$ W is a subspace of U.

Kernal of a Homomorphism

Definition: *Let f be a homomorphism of a vector space U(F) into a vector space V(F). The kernal W of f is defined as*

W = {$\alpha \in U$: $f(\alpha) = 0'$ where 0' is the zero vector of V}

Thus, the kernal W of f is subset of U consisting of those elements of U which are mapped under f onto the zero vector of V. Since f(0), 0', therefore at least $0 \in W$. Thus W is not empty.

2.14 ISOMORPHISM OF VECTOR SPACE

Definition: *Let U(F) and V(F) be two vector spaces. Then a mapping $f: U \to V$ is called an isomorphism of U onto V if*

(i) f is one-one,

(ii) f is onto,

(iii) $f(a\alpha + b\beta) = af(\alpha) + bf(\beta)\ \forall\ \alpha, \beta \in F,\ \forall\ \alpha, \beta \in U.$

Also then the two vector space U and V are said to be isomorphic and symbolically we write $U(F) \cong V(F)$.

The vector space V(F) is also called the isomorphic image of the vector space U(F).

If f is a homomorphism of U(F) into V(F), then f will become an isomorphism of U into V if f is one-one. Also in addition if f is onto V, then f will become a isomorphism of U onto V.

Isomorphism of Finite Dimensional Vector Spaces

Theorem 1:

Two finite dimensional vector spaces over the same field are isomorphic if and only if they are of the same dimension.

Proof:

First suppose that U(F) and V(F) are two finite dimensional vector spaces each of dimension n.

Then to prove that $U(F) \cong V(F)$.

Let the sets of vectors $\{\alpha_1, \alpha_2, ..., \alpha_n\}$ and $\{\beta_1, \beta_2, ..., \beta_n\}$ be the bases of U and V respectively.

Any vector $\alpha \in U$ can be uniquely expressed as

$$\alpha = a_1\alpha_1 + a_2\alpha_2 + ... + a_n\alpha_n.$$

Let $f: U \to V$ be defined by

$$f(\alpha) = a_1\beta_1 + a_2\beta_2 + ... + a_n\beta_n.$$

Since in the expression for α as a linear combination of $\alpha_1, \alpha_2, ..., \alpha_n$ the scalars $\alpha_1, \alpha_2, ..., \alpha_n$ are unique, therefore the mapping f is well-defined *i.e.*, $f(\alpha)$ is a unique element of V.

f is one-one. We have

$$f(a_1\alpha_1 + a_2\alpha_2 + ... + a_n\alpha_n) = f(b_1\alpha_1 + b_2\alpha_2 + ... + b_n\alpha_n)$$

$\Rightarrow a_1\beta_1 + a_2\beta_2 + ... + a_n\beta_n = b_1\beta_1 + b_2\beta_2 + ... + b_n\beta_n$

$\Rightarrow (a_1 - b_1)\ \beta_1 + ... + (a_n - b_n)\ b_n = 0'$ (zero vector of V)

$\Rightarrow a_1 - b_1 = 0,\ a_2 - b_2 = 0, ..., a_n - b_n = 0$ because

$\beta_1, \beta_2, ..., \beta_n$ are linearly independent

$\Rightarrow a_1 = b_1,\ a_2 = b_2, ..., a_n = b_n$

$\Rightarrow a_1\alpha_1 + a_2\alpha_2 + ... + a_n\alpha_n + b_1\alpha_1 + b_2\alpha_2 + ... + b_n\alpha_n$.

$\therefore$ f is one-one.

f is onto V. If $a_1\beta_1 + a_2\beta_2 + ... + a_n\beta_n$ is any element of V, then $\exists$ an element $a_1\alpha_1 + a_2\alpha_2 + ... + a_n\alpha_n \in U$ such that

$$f(a_1\alpha_1 + a_2\alpha_2 + ... + a_n\alpha_n) = a_1\beta_1 + a_2\beta_2 + ... + a_n\beta_n.$$

$\therefore$ f is onto V.

f is a linear transformation. We have

$f\ [a\ (a_1\alpha_1 + a_2\alpha_2 + ...\ a_n\alpha_n) + b\ (b_1\alpha_1 + b_2\alpha_2 + ... + b_n\alpha_n)$

$= f\ [(aa_1 + bb_1)\ \alpha_1 + (aa_2 + bb_2)\ \alpha_2 + ... + (aa_n + bb_n)\ \alpha_n)]$

$= (aa_1 + bb_1)\ \beta_1 + (aa_2 + bb_2)\ \beta_2 + ... + (aa_n + bb_n)\ \beta_n$

$= a\ (a_1\beta_1 + a_2\beta_2 + ... + a_n\beta_n) + b\ (b_1\beta_1 + b_2\beta_2 + ... + b_n\beta_n)$

$= af\ (a_1\alpha_1 + a_2\alpha_2 + ... + a_n\alpha_n) + bf\ (b_1\alpha_1 + b_2\alpha_2 + ... + b_n\alpha_n).$

$\therefore$ f is a linear transformation.

Hence f is an isomorphism of U onto V. $\therefore$ $U \cong V$.

Conversely, let U (F) and V(F) be two isomorphic finite dimensional vector spaces. Then to prove that dim U = dim V. Let dim U = n. Let $S = \{\alpha_1, \alpha_2, ..., \alpha_n)$ be a basis of U. If f is an isomorphism of U onto V, we shall show that $S' = \{f(\alpha_1), f(\alpha_2), ..., f(\alpha_n)\}$ is a basis of V. Then V will also be of dimension n.

First we show that S' is linearly independent.

Let $a_1\ f(\alpha_1) + a_2\ f(\alpha_2) + ... + a_n\ (f\ (\alpha_n) = 0$, (zero vector of V)

$\Rightarrow f\ (a_1\alpha_1 + a_2\alpha_2 + ... + a_n\alpha_n) = 0'$ [$\because$ f is linear transformation]

$\Rightarrow a_1\alpha_1 + a_2\alpha_2 + ... + a_n\alpha_n = 0$ [$\because$ f is one-one and f(0) = 0' where 0 is zero vector of U]

$\Rightarrow a_1 = 0,\ a_2 = 0, ..., a_n = 0$ since $\alpha_1, \alpha_2, ..., \alpha_n$ are linearly independent.

$\therefore$ S' is linearly independent.

Now to prove that L(S') = V. For this we shall prove that any vector $\beta \in V$ can be expressed as a linear combination of the vectors of the set S'. Since f is onto V, therefore $\beta \in V \Rightarrow$ there exists $\alpha \in U$ such that $f(\alpha) = \beta$.

Let $\alpha = c_1\alpha_1 + c_2\alpha_2 + ... + c_n\alpha_n$.

Then $\beta = f(\alpha) = f(c_1\alpha_1 + c_2\alpha_2 + ... + c_n\alpha_n)$

$= c_1 f(\alpha_1) + c_2 f(\alpha_2) + ... + c_n f(\alpha_n)$.

Thus β is a linear combination of the vectors of S'.

Hence V = L(S'). Therefore S' is a basis of V. Since S' contains n vectors, therefore dim V = n.

Note: While proving the converse, we have proved that *if f is an isomorphism of U onto V then f maps a basis of U onto a basis of V.*

Theorem 2:

Every n-dimensional vector space V(F) is isomorphic to $V_n(F)$.

Proof:

Let $\{\alpha_1, \alpha_2,..., \alpha_n\}$ be any basis of V(F). Then every vector $\alpha \in V$ can be uniquely expressed as

$$\alpha = a_1\alpha_1 + a_2\alpha_2 + ... + a_n\alpha_n,\ a_i \in F.$$

The ordered n-tuple $(a_1, a_2,..., a_n) \in V_n(F)$.

Let $f =: V(F) \rightarrow V_n(F)$ be defined by $f(\alpha) = (a_1, a_2,..., a_n)$.

Since in the expression of α as a linear combination of $\alpha_1, \alpha_2,..., \alpha_n$ the scalars $a_1, a_2,..., a_n$ are unique, therefore $f(\alpha)$ is a unique element of $V_n(F)$ and thus the mapping f is well-defined.

f is one-one. Let $\alpha = a_1\alpha_1 + a_2\alpha_2 + \quad + a_n\alpha_n$ and

$\beta = b_1\alpha_1 + b_2\alpha_2 + ... + b_n\alpha_n$ be any two elements of V. We have

$$f(\alpha) = f(\beta)$$

$$\Rightarrow \quad f(a_1\alpha_1 + a_2\alpha_2 + ... + a_n\alpha_n) = f(b_1\alpha_1 + b_2\alpha_2 + ... + b_n\alpha_n)$$

$$\Rightarrow \quad (a_1, a_2, ..., a_n) = (b_1, b_2,..., b_n)$$

$$\Rightarrow \quad a_1 = b_1, a_2 = b_2,..., a_n = b_n \Rightarrow \alpha = \beta.$$

$\therefore$ f is one-one.

f is onto $V_n(F)$. Let $(a_1, a_2,..., a_n)$ be any element of $V_n(F)$. Then there exists an element $a_1\alpha_1 + a_2\alpha_2 + ... + a_n\alpha_n \in V(F)$ such that $f(a_1\alpha_1 + a_2\alpha_2 + ... + a_n\alpha_n) = (a_1, a_2,..., a_n)$.

$\therefore$ f is onto V_n (F).

f is a linear transformation. If a, b $\in$ F and $\alpha, \beta \in$ V(F), we have f $(a\alpha + b\beta)$

$= f\,[a\,(a_1\alpha_1 + a_2\alpha_2 + ... + a_n\alpha_n) + b\,(b_1\alpha_1 + b_2\alpha_2 + .. + b_n\alpha_n)]$

$= f\,[(aa_1 + bb_1)\,\alpha_1 + (aa_2 + bb_2)\,\alpha_2 + ... + (aa_n + bb_n)\,\alpha_n]$

$= (aa_1 + bb_1, aa_2 + bb_2,..., aa_n + bb_n)$

$= (aa_1, aa_2,..., aa_n) + (bb_1, bb_2,..., bb_n)$

$= a\,(a_1, a_2,..., a_n) + b\,(b_1, b_2,..., bb_n)$

$= af\,(a_1\alpha_1 + a_2\alpha_2 + ... + a_n\alpha_n) + bf\,(b_1\alpha_1 + b_2\alpha_2 + ... + b_n\alpha_n)$

$= af\,(\alpha) + bf\,(\beta)$.

$\therefore$ f is an linear transformation.

$\therefore$ f is an isomorphism.

of V(F) onto V_n(F)

Hence V(F) $\cong V_n$(F).

Example 1:

If V is a finite dimensional vector space and f is an isomorphism of V into V, prove that f must map V onto V.

Solution:

Let V (F) be a finite dimensional vector space of dimension n. Let f be an isomorphism of V into V *i.e.* f is a linear transformation and f is one-one. To prove that f is onto V.

Let S = $\{\alpha_1, \alpha_2, ..., \alpha_n\}$ be a basis of V. We shall first prove that

S' = $\{f\,(\alpha_1), f\,(\alpha_2),..., f\,(\alpha_n)\}$ is also a basis of V. We claim that S' is linearly independent. The proof is as follows:

Let $a_1\, f\,(\alpha_1) + a_2 f\,(\alpha_2) + ... + a_n\, f\,(\alpha_n) = 0$ (zero vector of V)

$\Rightarrow \quad f\,(a_1\alpha_1 + a_2\alpha_2 + ... + a_n\alpha_n) = 0$ [$\because$ f is a linear transformation]

$\Rightarrow \quad a_1a_1 + a_2a_2 + ... + a_n\alpha_n = 0$ [$\because$ f is one-one and f (0) = 0]

$\Rightarrow \quad a_1 = 0, a_2 = 0,..., a_n = 0$

since $\alpha_1, \alpha_2,..., \alpha_n$ are linearly independent.

$\therefore$ S' is linearly independent.

Now V is of dimension n and S' is a linearly independent subset of V containing n vectors. Therefore S' must be a basis of V.

Therefore each vector in V can be expressed as a linear combination of the vectors belonging to S'.

Now we shall show that f is onto V. Let α be any element of V. Then there exist scalars $c_1, c_2, \ldots c_n$ such that

$$\alpha = c_1 f(\alpha_1) + c_2 f(\alpha_2) + \ldots + c_n f(\alpha_n)$$
$$= f(c_1\alpha_1 + c_2\alpha_2 + \ldots + c_n\alpha_n).$$

Now $c_1\alpha_1 + c_2\alpha_2 + \ldots + c_n\alpha_n \in V$ and the f image of this element is a. Therefore f is onto V. Hence f is an isomorphism of V onto V.

Example 2:

If V is finite dimensional and f is a homomorphism of V into itself which is not onto prove that there is some $\alpha \neq 0$ in V such that $f(\alpha) = 0$.

Solution:

If f is a homomorphism of V into itself, then f (0) = 0. Suppose there is no non-zero vector a in V such that $f(\alpha) = 0$.

Then f is one-one. Because

$$f(\beta) = f(\gamma) \Rightarrow f(\beta) - f(\gamma) = 0$$
$$\Rightarrow f(\beta - \gamma) = 0 \qquad [\because \text{ f is linear transformation}]$$
$$\Rightarrow \beta - \gamma = 0 \Rightarrow \beta = \gamma.$$

Now V is finite dimensional and f is a linear transformation of V into itself. Since f is one-one, therefore f must be onto V. But it is given that f is not onto. Therefore our assumption is wrong. Hence there will be a non-zero vector α in V such that

$$f(\alpha) = 0.$$

Example 3:

The mapping $f: V_3(F) \rightarrow V_2(F)$ defined by

$$f(a_1, a_2, a_3) = (a_1, a_2)$$

is a homomorphism of V_3 (F) onto V_2 (F). What is the kernel of this homomorphism ?

Solution:

Let $\alpha = (a_1, a_2, a_3)$ and $\beta - (b_1, b_2, b_3)$ be any two elements of V_3 (F). Also let a, b be any two elements of F. We have

$$f(a\alpha + b\beta) = f[a(a_1, a_2, a_3) + b(b_1, b_2, b_3)]$$
$$= f[(aa_1 + bb_1, aa_2 + bb_2, aa_2 + bb_3)] = (aa_1 + bb_1, aa_2 + bb_2)$$

$= a\,(a_1, a_2) + b\,(b_1, b_2) = af\,(a_1, a_2, a_2) + bf\,(b_1, b_2, b_3)$

$= af\,(\alpha) + bf\,(\beta)$.

$\therefore$ f is a linear transformation.

To show that f is onto V_2 (F). Let (a_1, a_2) be any element of V_2 (F). Then $(a_1, a_2, 0) \in V_2$ (F) and we have $f\,(a_1, a_2, 0) = (a_1, a_2, 0) = (a_1, a_2)$. Therefore f is onto V_2 (F).

Therefore f is a homomorphism of V_3 (F) onto V_2 (F). If W is the kernel of this homomorphism then $W = \{(0, 0, a) : a \in F\}$.

We have $\forall\, a \in F$, $f\,(0, 0, a) = (0, 0)$ = the zero vector of V_2 (F).

Also if $f\,(a_1, a_2, a_3) = (0, 0)$ then $f\,(a_1, a_2, a_3) = (a_1, a_2) = (0, 0)$ implies $a_1 = 0$, $a_2 = 0$. Therefore $(a_1, a_2, a_3) \in W$.

Hence W is the kernel of f.

Example 4:

If f: U → V is an isomorphism of the vector space U into the vector space V, then a set of vectors $\{f(\alpha_1), f(\alpha_2), \ldots, f(\alpha_r)\}$ *is linearly independent if and only if the set* $\{\alpha_1, \alpha_2, \ldots, \alpha_r\}$ *is linearly independent.*

Solution:

U(F) and V(F) are two vector spaces over the same field F and f is isomorphism of U into V *i.e.*,

f: U → V such that

f is 1-1 and $f\,(a\alpha + b\beta) = af\,(\alpha) + bf\,(\beta)$

$\forall\, a, b \in F$ and $\forall\, \alpha, \beta, \in U$.

Let $\{\alpha_1, \alpha_2, \ldots, \alpha_r\}$ be a subset of U. First suppose that the vectors $\alpha_1, \alpha_2, \ldots, \alpha_r$ are linearly independent. Then to show that the vectors $f\,(\alpha_1)$, $f(\alpha_2), \ldots, f\,(\alpha_r)$ re also linearly independent.

We have

$a_1 f\,(\alpha_1) + a_2\,f\,(\alpha_2) + \ldots + a_r\,f\,(\alpha_r) = 0,$ where $a_1, a_2, \ldots, a_r \in F$

$\Rightarrow f\,(a_1\alpha_1 + a_2\alpha_2 + \ldots + a_r\alpha_r) = 0$ [$\because$ f is a linear transformation]

$\Rightarrow f\,(a_1\alpha_1 + a_2\alpha_2 + \ldots + a_r\alpha_r) = f\,(0)$ [$\because f\,(0) = 0$]

$\Rightarrow a_1\alpha_1 + a_2\alpha_2 + \ldots + a_r\alpha_r = 0$ [$\because$ f is 1-1]

$\Rightarrow a_1 = 0, a_2 = 0, \ldots, \alpha_r = 0$ since the vectors $\alpha_1, \alpha_2, \ldots, \alpha_r$ are linearly independent.

Hence the vectors $f\,(\alpha_1), f\,(\alpha_2), \ldots, f\,(\alpha_r)$ are also linearly independent.

Conversely suppose that the vectors f (α_1), f (α_2),..., f (α_r) also linearly independent.

Conversely suppose that the vectors f (α_1) f (α_2),..., f (α_r) linearly independent. Then to show that the vectors $\alpha_1, \alpha_2, ..., \alpha_r$ are also linearly independent.

We have

$$a_1\alpha_1 + a_2\alpha_2 + ... a_r\alpha_r = 0, \text{ where } a_1, a_2, ..., a_r \in F$$

$$\Rightarrow \quad f(a_1\alpha_1 + a_2\alpha_2 + ... + a_r\alpha_r) = f(0)$$

$$\Rightarrow \quad a_1 f(\alpha_1) + a_2 f(\alpha_2) + ... + a_r f(\alpha_r) = 0$$

[$\because$ f is a linear transformation]

$$\Rightarrow \quad a_1 = 0, a_2 = 0, ..., a_r = 0$$

since the vectors f (α_1), f (α_2),..., f (α_r) linearly independent.

Hence the vectors $\alpha_1, \alpha_2, ..., \alpha_r$ are also linearly independent.

Example 5:

If V is finite dimensional and f is a homomorphism of V onto V prove that f must be one-one, and so, an isomorphism.

Solution:

Let V (F) be a finite dimensional vector space of dimension n. Let f be a homomorphism of V onto V *i.e.*, f is a linear transformation and f is onto V. To prove that f is one-one.

Let S = $\{\alpha_1, \alpha_2, ..., \alpha_n\}$ be a basis of V. We shall first prove that S' = {f (α_1), f (α_2),..., f (α_n) is also a basis of V. We claim that L (S') = V. The proof is as follows:

Let α be any element of V. We shall show that α can be expressed as a linear combination of f (α_1), f (α_2), ... f (α_n). Since f is onto V, therefore $\alpha \in$ V implies that there exists $\beta \in$ V such that f $(\beta) = \alpha$. Now β can be expressed as a linear combination of $\alpha_1, \alpha_2, ..., \alpha_n$. Let

$$\beta = a_1\alpha_1 + a_2\alpha_2 + ... + a_n\alpha_n.$$

Then
$$\alpha = f(\beta) = f(a_1\alpha_1 + a_2\alpha_2 + ... + a_n\alpha_n)$$
$$= a_1 f(\alpha_1) + a_2 f(\alpha_2) + \quad + a_n f(\alpha_n).$$

Thus α has been expressed as a linear combination of

$$f(\alpha_1), f(\alpha_2), ..., f(\alpha_n).$$

Therefore L(S') = V.

Since is of dimension n and S' is a subset of V containing n vectors and L(S') = V, therefore S' must be a basis of V. Therefore each vector in V can be expressed as a linear combination of the vectors belonging to S' and S' is linearly independent.

Now, we shall that f is one-one. Let γ, δ be any two elements of V such that $\gamma = c_1\alpha_1 + c_2\alpha_2 + ... + c_n\alpha_n$, $\delta = d_1\alpha_1 + d_2\alpha_2 + ... + d_n\alpha_n$.

We have $f(\gamma) = f(\delta)$

$\Rightarrow \quad f(c_1\alpha_1 + c_2\alpha_2 + ... + c_n\alpha_n) = f(d_1\alpha_1 + d_2\alpha_2 + ... + d_n\alpha_n)$

$\Rightarrow \quad c_1 f(\alpha_1) + c_2 f(\alpha_2) + ... + c_n f(\alpha_n) = d_1 f(\alpha_1) + d_2 f(\alpha_2) + ... + d_n f(\alpha_n)$

$\Rightarrow \quad (c_1 - d_2) f(\alpha_1) + (c_2 - d_2) f(\alpha_2) + ... + (c_n - d_n) f(\alpha_n) = 0$

$\Rightarrow \quad c_1 - d_1 = 0, c_2 - d_2 = 0, ..., c_n - d_n = 0$ since $f(\alpha_1), f(\alpha_2), ..., f(\alpha_n)$ are linearly independent

$\Rightarrow \quad c_1 = d_1, c_2 = d_2, ..., c_n = d_n \Rightarrow \gamma = \delta.$

$\therefore$ f is one-one.

Hence f is an isomorphism of V onto V.

Example 6:

Let f be a linear transformation from a vector space U into a vector space V. If S is a subspace of U, prove that f (S) will be a subspace of V.

Solution:

U (F) and V (F) are two vector spaces over the same field F. The mapping f is a linear transformation of U into V *i.e.*,

$f: U \to V$ such that

$f(a\alpha + b\beta) = af(\alpha) = bf(\beta)\ \forall\ a, b \in F$ and $\forall\ \alpha, \beta \in U$.

Let S be a subspace of U. Then to prove that f (s) is a subspace of V.

Let $a, b \in F$ and $f(\alpha), f(\beta) \in f(S)$ where $\alpha, \beta \in S$.

Since S is a subspace of U, therefore

$$a, b \in F \text{ and } \alpha, \beta \in S \Rightarrow a\alpha + b\beta \in S$$

$\Rightarrow \quad f(a\alpha + b\beta) \in f(S)$

$\Rightarrow \quad af(\alpha) + bf(\beta) \in f(S). \quad [\because f(a\alpha + b\beta) = af(\alpha) + bf(\beta)].$

Thus $a, b \in F$ and $f(\alpha), f(\beta) \in f(S)$

$\Rightarrow \quad af(\alpha) + bf(\beta) \in f(S).$

Hence f (S) is a subspace of V.

Example 7:

Define linear transformation of a vector space V (F) into a vector space W(F). Show that the mapping

$$T: (a, b) \rightarrow (a + 2, b + 3)$$

of V_2 **(R)** *into itself is not a linear transformation.*

Solution:

Linear Transformation. Definition. Let V(F) and W(F) be two vector spaces over the same field F. A mapping

$$T: V \rightarrow W$$

is called a linear transformation of V into W if

$$T(a\alpha + b\beta) = a\,T(\alpha) + b\,T(\beta)\ \forall\ a, b, \in F \text{ and } \forall\ \alpha, \beta \in V.$$

Now to show that the mapping

$$T: (a, b) \rightarrow (a + 2, b + 3)$$

of V_2 (**R**) into itself is not a linear transformation.

Take $\alpha = (1, 2)$ and $\beta = (1, 3)$ as two vectors of V_2(**R**) and $a = 1$, $b = 1$ as two elements of the field **R**.

Then $a\alpha + b\beta = 1\,(1, 2) + 1\,(1, 3) = (1, 2) + (1, 3)$

$= (2, 5)$.

By the definition of the mapping T, we have

$$T(a\alpha + b\beta) = T(2, 5) = (2 + 2, 5 + 3) = (4, 8). \quad ...(1)$$

Also $T(\alpha) = T(1, 2) = (1 + 2, 2 + 3) = (3, 5)$

and $T(\beta) = T(1, 3) = (1 + 2, 3 + 3) = (3, 6)$.

$\therefore aT(\alpha) + bT(\beta) = 1\,(3, 5) + 1\,(3, 6)$

$= (3, 5) + (3, 6) = (6, 11). \quad ...(2)$

From (1) and (2), we see that

$$T(a\alpha + b\beta) \neq aT(\alpha) + bT(\beta).$$

Hence T is not a linear transformation of V_2 (**R**) into itself.

2.15 DIRECT SUM OF SPACES

Vector Space as Direct Sum of Subspaces

Definition: *Let V(F) be a vector space and let* $W_1, W_2, \ldots, W_m$ *be subspaces of V. Then V is said to be the direct sum of* $W_1, W_2, \ldots, W_m$ *if every element* $a \in V$ *can be written in one and only one way as* $a = a_1 + a_2 + \ldots + a_m$ *where*

$a_1 \in W_1, a_2 \in W_2, \ldots, a_m \in W_m.$

If a vector space V(F) is a direct sum of its two subspaces W_1 and W_2 then we should have not only $V = W_1 + W_2$ but also that each vector V can be uniquely expressed as sum of an element of W_1 and an element of W_2. Symbolically the direct sum is represented by the notation $V = W_1 \oplus W_2$.

Disjoint Subspaces

Definition: *Two subspaces W_1 and W_2 of the vector space V(F) are said to be disjoint if their intersection is the zero subspace i.e., if $W_1 \cap W_2 = \{0\}$.*

Theorem:

The necessary and sufficient condition for a vector space V(F) to be direct sum of its two subspaces W_1 and W_2 are that

(i) $V = W_1 + W_2$

and *(ii)* $W_1 \cap W_2 = \{0\}$

i.e., *W_1 and W_2 are disjoint.*

Proof:

Let V be a direct sum of its two subspaces W_1 and W_2. Then each element of V is expressible as sum of an element of W_1 and an element of W_2. Therefore we have $V = W_1 + W_2$.

Let, if possible $0 \neq \alpha \in W_1 \cap W_2$. Then $\alpha \in W_1, \alpha \in W_2$. Also $\alpha \in V$ and we can write

$$\alpha = 0 + \alpha \text{ where } 0 \in W_1, \alpha \in W_2$$

and $$\alpha = \alpha + 0 \text{ where } \alpha \in W_1, 0 \in W_2.$$

Thus, $\alpha \in V$ can be expressed in at least two different ways as sum of an element of W_1 and an element of W_2. This contradicts the fact that V is a direct sum of W_1 and W_2. Hence 0 is the only vector common to both W_1 and W_2 *i.e.*, $W_1 \cap W_2 = \{0\}$. Thus, the conditions are necessary.

The conditions are sufficient. Let $V = W_1 + W_2$ and $W_1 \cap W_2 = \{0\}$. Then to show that V is direct sum of W_1 and W_2.

$V = W_1 + W_2 \Rightarrow$ that each element V can be expressed as sum of an element of W_1 and an element of W_2. Now to show that this expression is unique.

Let, if possible,

$$\alpha = \alpha_1 + \alpha_2, \alpha \in V, \alpha_1 \in W_1, \alpha_2 \in W_2$$

and $$\alpha = \beta_1 + \beta_2, \beta_1 \in W_1, \beta_2 \in W_2.$$

Then to show that $\alpha_1 = \beta_1$ and $\alpha_2 - \beta_2$.

We have $\quad \alpha_1 + \alpha_2 = \beta_1 + \beta_2$

$\Rightarrow \quad \alpha_1 - \beta_1 = \beta_2 - \alpha_2$.

Since W_1 is a subspace, therefore

$$\alpha_1 \in W_1, \beta_1 \in W_1$$

$$\Rightarrow \quad \alpha_1 - \beta_1 \in W_1.$$

Similarly $\beta_2 - \alpha_2 \in W_2$.

$\therefore \alpha_1 - \beta_1 = \beta_2 - \alpha_2 \in W_1 \cap W_2$.

But 0 is the only vector which belongs to $W_1 \cap W_2$. Therefore

$$\alpha_1 - \beta_1 = 0 \Rightarrow \alpha_1 = \beta_1.$$

Also $\quad \beta_2 - \alpha_2 = 0 \Rightarrow \alpha_2 = \beta_2$

Thus each vector $\alpha \in V$ is uniquely expressible as sum of an element W_1 and an element of W_2. Hence $V = W_1 \oplus W_2$.

Dimension of a Direct Sum

Theorem:

If a finite dimensional vector space V (F) is a direct sum of its two subspaces W_l and W_2, then

$$dim\ V = dim\ W_l + dim\ W_2.$$

Proof:

Let dim $W_1 = m$ and dim $W_2 = l$. Also let the sets of vectors $S_1 = \{\alpha_1, \alpha_2, ..., \alpha_m\}$ and $S_2 = \{\beta_1, \beta_2, ..., \beta_1\}$ be some bases of W_1 and W_2 respectively.

We have dim W_1 + dim $W_2 = m + l$.

In order to prove that dim V = dim W_1 + dim W_2, we should therefore prove that dim $V = m + l$. We claim that the set

$$S = S_1 \cup S_2 = \{\alpha_1, \alpha_2, ..., \alpha_m, \beta_1, \beta_2, ..., \beta_l\}$$

is a basis of V.

First we show that the set S is linearly independent. Let

$$a_1\alpha_1 + a_2\alpha_2 + ... + a_m\alpha_m + b_1\beta_2 + b_2\beta_2 + ... + b_l\beta_l - 0$$

$$\rightarrow \quad a_1\alpha_1 + a_2\alpha_2 + ... + a_m\alpha_m = -(b_1\beta_1 + b_2\beta_2 + ... + b_l\beta_l)$$

Now $\quad a_1\alpha_1 + a_2\alpha_2 + ... + a_m\alpha_m \in W_l$

and $\quad -(b_1\beta_1 + b_2\beta_2 + ... + b_1\beta_1) \in W_2$. Therefore

$$a_1\alpha_1 + a_2\alpha_2 + ... + a_m\alpha_m \in W_1 \cap W_2$$

and $\quad -(b_1\beta_1 + b_2\beta_2 + ... + b_l\beta_l) \in W_1 \cap W_2$.

But V is the direct sum of W_1 and W_2. Therefore 0 is the only vector belonging to $W_1 \cap W_2$. Then we have

$$a_1\alpha_1 + a_2\alpha_2 + ... + a_m\alpha_m = 0,\ b_1\beta_1 + b_2\beta_2 + ... + b_l\beta_l = 0.$$

Since both the sets $\{\alpha_1, \alpha_2,..., \alpha_m\}$ and $\{\beta_1, \beta_2,..., \beta_l\}$ are linearly independent, therefore we have

$$a_1 = 0, a_2 = 0,..., a_m = 0, b_1 = 0, b_2 = 0,..., b_l = 0.$$

Therefore S is linearly independent.

Now we shall show that L(S) = V. Let α be any element of V. Then

α = an element of W_1 + an element of W_2

= a linear combination of S_1 + a linear combination of S_2

= a linear combination of S.

$\therefore$ L(S) = V.

$\therefore$ S is a basis of V. Therefore dim V = m + l.

Hence the theorem.

Complementary Subspaces

Definition. *Let V(F) be a vector space and W_1, W_2 be two subspace of V. Then the subspace W_2 is called a complement of W_1 in V if V is the direct sum of W_1 and W_2.*

Existence of Complementary Subspaces

Theorem:

Corresponding to each subspace W_1 of a finite dimensional vector space V(F), there exists a subspace W_2 such that V is the direct sum of W_1 and W_2.

Proof:

Let dim W_1 = m. Let the set $S_1 = \{\alpha_1, \alpha_2,..., \alpha_m\}$ be a basis of W_1. Since S_1 is a linearly independent subset of V, therefore S_1 can be extended to form a basis of V. Let the set $S = \{\alpha_1, \alpha_2,..., \alpha_m, \beta_1, \beta_2,..., \beta_l\}$ be a basis of V. Let W_2 be the sub-space of V generated by the set $S_2 = \{\beta_1, \beta_2,..., \beta_l\}$.

We shall prove that V is the direct sum of W_1 and W_2. For this we shall prove that $V = W_1 + W_2$ and $W_1 \cap W_2 = \{0\}$.

Let α be any element of V. Then we can express

α = a linear combination of S

$= $ a linear combination of S_1 + a linear combination of S_2

$= $ an element of W_1 + an element of W_2.

$\therefore V = W_1 + W_2$.

Again let $\beta \in W_1 \cap W_2$. The β can be expressed as a linear combination of S_1 and also a linear combination of S_2. So we have

$$\beta = a_1\alpha_1 + a_2\alpha_2 + \ldots + a_m\alpha_m = b_1\beta_1 + b_2\beta_2 + \ldots b_l\beta_l.$$

$$\therefore a_1\alpha_1 + a_2\alpha_2 + \ldots + a_m\alpha_m - b_1\beta_1 - b_2\beta_2 - \ldots b_l\beta_l = 0$$

$\Rightarrow a_1 = 0, a_2 = 0, \ldots, a_m = 0, b_1 = 0, b_2 = 0, \ldots, b_l = 0$ since $\alpha_1, \alpha_2, \ldots, \alpha_m, \beta_1, \beta_2, \ldots, \beta_l$ are linearly independent.

$\therefore \beta = 0$ (zero vector).

Thus $W_1 \cap W_2 = \{0)$ Hence V is the direct sum of W_1 and W_2.

2.16 COORDINATES

Let V (F) be a finite dimensional vector space. Let $B = \{\alpha_1, \alpha_2, \ldots, \alpha_m\}$ be an ordered basis for V. By an ordered basis we mean that the vectors of B have been enumerated in some well-defined way *i.e.*, the vectors occupying the first, second,..., n^{th} places in the set B are fixed.

Let $\alpha \in V$. Then there exists a unique n-tuple $(x_1, x_2, \ldots, x_n)$ of scalars such that $\alpha = x_1\alpha_1 + x_2\alpha_2 + \ldots + x_n\alpha_n = \sum_{l=1}^{n} x_l a_l$.

The n-tuple $(x_1, x_2, \ldots, x_n)$ is called the n-tuple of coordinates of α relative to the ordered basis B. The scalar x_l is called the ith coordinate of α relative to the ordered basis B.

It should be noted that for the same basis set B, the coordinates of the vector α are unique only with respect to a particular ordering of B. The basis set B can be ordered in several ways. The coordinates of α may change with change in ordering of B.

Example:

Show that the set $S = \{(1, 0, 0), (1, 1, 0), (1, 1, 1)\}$ is a basis of $\mathbf{R}^3$ **(R)** *where* **R** *is the field of real numbers. Hence find the coordinates of the vector (a, b, c) with respect to the above basis.*

Solution:

The dimension of the vector space $\mathbf{R}^3(\mathbf{R})$ is 3. If the set S is linearly independent, then S will form a basis of $\mathbf{R}^3(\mathbf{R})$ because S contains 3 vectors. Let x, y, z be scalars in R such that

$$x\,(1, 0, 0) + y\,(1, 1, 0) + z\,(1, 1, 1) = 0 = (0, 0, 0)$$

$\Rightarrow$ $(x + y + z, y + z, z) = (0, 0, 0)$

$\Rightarrow$ $x + y + z = 0, y + z = 0, z = 0$

$\Rightarrow$ $x = 0, y = 0, z = 0$

$\Rightarrow$ the set S is linearly independent.

$\therefore$ S is a basis of $\mathbf{R}^3(\mathbf{R})$.

Now to find the coordinates of (a, b, c) with respect to the ordered basis S. Let p, q, r be scalars in **R** such that

$$(a, b, c) = p\,(1, 0, 0) + q\,(1, 1, 0) + r\,(1, 1, 1)$$

$\Rightarrow$ $(a, b, c) = (p + q + r, q + r, r)$

$\Rightarrow$ $p + q + r = a, q + r = b, r = c$

$\Rightarrow$ $r = c, q = b - c, p = a - b.$

Hence the coordinates of the vector (a, b, c) are (p, q, r)

i.e., $(a - b, b - c, c)$.

3

RINGS (CONTINUED)

3.1 DIVISIBILITY OF POLYNOMIALS OVER A FIELD

Suppose F is a field. Then F [x] is an integral domain. If $a(x) \neq 0$ and f(x) are elements of F [x], then a(x) is a *divisor (or factor)* of f (x) if and only if there is a polynomial b (x) in F [x] such that f (x) = a (x) b (x). Symbolically we write a (x) | f (x).

Unit

A *unit* is an element of F [x] which has multiplicative inverse. All the polynomials of zero degree belonging to F [x] are units of F [x]. Thus, the non-zero elements of F are the only units of F [x].

If f [x] and g [x] are polynomials in F [x], then we call f [x] and g (x) associates if f (x) = c g (x) for some $0 \neq c \in F$. It can be easily proved that two non-zero polynomials f (x) and g (x) in F [x] are associates if and only if f (x) | g (x) and g (x) | f (x).

If f (x) is any non-zero polynomial in F [x], then f (x) is always divisible by its associates and by all units of F [x]. These divisors of f (x) are called its *improper divisor.* All other divisors of f (x), if there are any, R called its *proper divisors.*

Definition of an Irreducible Polynomial Over a Field

Let F be a field and f (x) be a non-zero and non-unit polynomial in F [x] i.e., f (x) be a polynomial of positive degree. Then f (x) is said to be irreducible over F (or prime) if it has no proper divisors in F [x]; f (x) is reducible over F if it has a proper divisor in F [x].

Thus, a positive degree polynomial f (x) in F [x] is irreducible over F if whenever f (x) = a (x) b (x) with a (x), b (x) $\in$ F [x] then one of a (x) or b (x) in is a unit in F [x] *i.e.*, has degree 0. Also f (x) is reducible over F if and only if we can find two polynomials a (x) and b (x) in F [x] such that f (x) = a (x) b (x) and none of a (x) and b (x) is a unit in F [x] *i.e.*, has degree 0.

Irreducibility depends on the field. The polynomial $x^2 - 2$ is irreducible over the field of rational numbers while it is reducible over the field of real numbers, since $x^2 - 2 = (x + \sqrt{2})(x - \sqrt{2})$.

The polynomial $x^2 + 1$ is irreducible over the field of real numbers while it is reducible over the field of complex numbers since $x^2 + 1 = (x + i)(x - i)$.

Monic Polynomials

Definition: *Let* $f(x) = a_0 + a_1 x + \ldots + a_n x^n$, *with* $a_n \neq 0$, *be a polynomial in F [x] over an arbitrary field F. If the leading coefficient* a_n *of f (x) is equal to 1, the unity element of F, then the polynomial f (x) will be called monic.*

The polynomial $2x - 3x^3$ over the field of real numbers is not monic since its leading coefficient is – 3. But the polynomial $x^3 - 3x + 4$ over the field of real numbers is monic since its leading coefficient is 1.

Greatest Common Divisor of Two Polynomials Over a Field

Definition: *Suppose F is any field. Let f (x) and g (x) be two elements of F [x]. A greatest common divisor of F (x) and g (x) is a non-zero polynomial d (x) such that*

(i) d (x) | f (x) and d (x) g (x)

(ii) If c (x) is a polynomial such that c (x) | f (x) and c (x) | g (x) then c (x) | d (x).

Relatively Prime Polynomials

Two polynomials f (x) and g (x) ∈ F [x] are said to be relatively prime if their greatest common divisor is 1, the unity element of F.

3.2 DIVISION ALGORITHM FOR POLYNOMIALS OVER A FIELD

Theorem:

Let f (x), g (x) ≠ 0 be any two polynomials of the polynomial domain F [x], over the field F. Then there exist uniquely two polynomials q (x) and r (x) in F [x] such that

$$f(x) = q(x)\, g(x) + r(x)$$

where either $r(x) = 0$ *or* $\deg r(x) < \deg g(x)$.

Proof:

Suppose

$$f(x) = a_0 + a_1x + a_2x^2 + \dots + a_mx^m,\ a_m \neq 0$$

and $$g(x) = b_0 + b_1x + b_2x^2 + \dots + b_nx^n,\ b_n \neq 0.$$

If degree m of f (x) is smaller than the degree n of g (x) or if f (x) =0, then we are nothing to prove. Because we can always write f (x) = 0 . g (x). So in this case q (x) = 0, r (x) = f (x) and we have either f (x) = 0 or deg f (x) < deg g (x).

Now let us assume that $m \geq n$. In this case we shall prove the theorem by induction on m. *i.e.*, degree of f (x).

If m = 0, then $m \geq n \Rightarrow n = 0$. Therefore f (x) and g (x) are both non-zero constant polynomials, $f(x) = a_0$, $a_0 \neq 0$, and $g(x) = b_0$, $b_0 \neq 0$. We have in this case

$$f(x) = a_0 = \left(a_0b_0^{-1}\right)b_0 + 0 = \left(a_0b_0^{-1}\right)g(x) + 0.$$

Thus, the theorem is true when m = 0 or when the degree of f (x) is less than 1.

We shall now assume that the theorem is true when f (x) is a polynomial of degree less than m and then we shall show that it is also true if f (x) is of degree m and then the proof will be complete by induction.

Let $$f_1(x) = f(x) - (a_m\ b_n^{-1})\ x^{m-n}\ g(x). \qquad \dots(1)$$

Obviously deg $f_1(x) < m$. Therefore by our assumed hypothesis, there exist polynomials s (x) and r (x) such that

$$f_1(x) = s(x)\ g(x) + r(x),$$

where r (x) = 0 or deg r (x) < deg g (x).

Now putting the value of $f_1(x)$ in (1)

$$s(x)\ g(x) + r(x) + f(x) - \left(a_m\ b_n^{-1}\right)\ x^{m-n}\ g(x)$$

$$\Rightarrow \quad f(x) = \left[\left(a_m\ b_n^{-1}\right)\ x^{m-n} + s(x)\right] g(x) + r(x).$$

If we write q (x) in place of $\left(a_m\ b_n^{-1}\right)\ x^{m-n} + s(x)$, we get

$$f(x) = q(x)\ g(x) + r(x)$$

where r (x) = 0 or deg r (x) < deg g (x).

This proves the existence of polynomials q (x) and r (x). Now to show that q (x) and r (x) are unique. Let us assume that

$$f(x) = q_1(x) g(x) + r_1(x) = q_2(x) g(x) + r_2(x).$$

Then $q_1(x) g(x) + r_1(x) = q_2(x) g(x) + r_2(x)$

$\Rightarrow$ $[q_1(x) - q_2(x) g(x) = r_2(x) - r_1(x).$ ...(2)

If $[q_1(x) - q_2(x)] \neq 0$, then $[q_1(x) - q_2(x)] g(x)$ cannot be equal to the zero polynomial because g (x) ' 0 and F [x] is without zero divisors. Also then the degree of $[q_1(x) - q_2(x)] g(x)$ is at least n, the degree of g (x). But $r_2(x) - r_1(x)$ is either equal to the zero polynomial or else its less than n because the degrees of $r_2(x)$ and $r_1(x)$ are both less than n. Hence the equality (2) among two polynomials holds only if $q_1(x) - q_2(x) = 0$ and $r_2(x) - r_1(x) = 0$

i.e., only if $q_1(x) = q_2(x)$ and $r_1(x) = r_2(x)$.

$\therefore$ the polynomials q (x) and r (x) are unique.

Definition: *In the division algorithm, the polynomial q (x) is called the* **quotient** *on dividing f (x) by g (x) and the polynomial r (x) is called the* **remainder.**

Theorem:

A polynomial domain F [x] over a field F is a principal ideal ring.

Proof:

Obviously F [x] is a commutative ring with unity and without zero divisors. Therefore F [x] is a principal ideal ring if every ideal in F [x] is a principal ideal.

Let S be an arbitrary ideal of F [x]. If S is the null ideal, then S = (0) *i.e.*, the ideal of F [x] generated by 0. Therefore S is a principal ideal, so let us suppose that S is not a null ideal. Then there exist non-zero polynomials f (x) in S. Let g (x) be a polynomial of lowest degree m belonging to S. We shall show that S is the principal ideal generated by g (x).

Let f (x) be any arbitrary member of S. By division algorithm there exist two polynomials $q(x) \in F[x]$, $r(x) \in F[x]$, such that

$f(x) = q(x) g(x) + r(x)$, where $r(x) = 0$ or $\deg r(x) < \deg g(x)$.

Since S is an ideal, therefore

$$q(x) \in F[x], g(x) \in S \Rightarrow q(x) g(x) \in S.$$

Also $f(x) \in S, q(x) g(x) \in S \Rightarrow f(x) - q(x) g(x) \in S.$

But $f(x) - q(x) g(x) = r(x)$. Therefore $r(x) \in S$.

Now either $r(x) = 0$ or $\deg r(x) < \deg g(x)$. But we have assumed that g (x) is a polynomial of lowest degree belonging to S. Hence deg r (x) cannot be less than deg (x). Therefore we must have r (x) = 0. Then f (x) = q (x) g (x). Thus g (x) ∈ S is such that f (x) ∈ S ⇒ f (x) = q (x) g (x) for some q (x) ∈ F [x]. Therefore S is a principal ideal of F [x] generated by g (x). Hence F [x] is a principal ideal ring.

Note: A polynomial ring over an arbitrary field is a principal ideal ring. But a polynomial ring over an arbitrary ring is not a principal ideal ring as is obvious from the following example.

Example:

Show that the polynomial ring **I** *[x] over the ring of integers is not a principal ideal ring.*

Solution:

To prove this statement we shall show that the ideal (2, x) of the ring I [x] generated by two elements 2 and x of **I** [x] is not a principal ideal. Let (2, x) be a principal ideal in **I** [x]. Then there will exist a non-zero element g (x) ∈ **I** [x] such that (2, x) = (g (x)).

Since 2 ∈ (g (x)) and x ∈ (g (x)) therefore there will exist elements φ (x) and ψ (x) belonging to **I** [x] such that

$$2 = \phi (x) g (x) \quad ...(1)$$

$$x = \psi (x) g (x). \quad ...(2)$$

From (1), we get 2x = [φ (x) g (x)] x and from (2), we get

$$2x = 2 \psi (x) g (x).$$

∴ 2 ψ (x) g (x) = x φ (x) g (x) [∴ **I** [x] is a commutative ring].

∴ 2 ψ (x) = x φ (x) since g (x) ≠ 0, and **I** [x] is without zero divisors.

Now 2 ψ (x) = x φ (x) implies that the coefficients of φ (x) must be even integers. Therefore φ (x) = 2h (x) where h (x) is some polynomial in **I** [x]. Putting this value of φ (x) in (1) we get

$$2 = 2h (x) g (x)$$

$$\Rightarrow \quad 1 = h (x) g (x).$$

Now 1 = h (x) g (x) ⇒ 1 ∈ (g (x)). Therefore element of **I** [x] will belong to (g (x)). Thus, we have **I** [x] = (g (x)) = (2, x). Therefore each element of **I** [x] will belong to (2, x). We shall show that 1 ∉ (2, x) and this contradiction will mean that (2, x) is not a principal ideal in **I** [x].

Now 1 ∈ (2, x) ⇒ we can write

$$1 = 2p (x) + xq (x)$$

where p (x) and q (x) are some elements in **I** [x].

Let $p(x) = a_0 + a_1 x + a_2x^2 + \ldots$ and $q(x) = b_0 + b_1x + b_2x^2 + \ldots$

Then $1 = 2(a_0 + a_1x + a_2x^2 + \ldots) + x(b_0 + b_1x + b_2x^2 + \ldots)$

$\Rightarrow 1 = 2a_0 + (2a_1 + b_0)x + (2a_2 + b_1)x^2 + \ldots$

This equality implies $1 = 2a_0$ where $a_0 \in \mathbf{I}$.

But for no integer a_0 we can have $1 = 2a_0$. Hence $1 \notin (2, x)$.

$\therefore$ (2, x) is not a principal ideal in **I** [x].

3.3 EUCLIDEAN ALGORITHM FOR POLYNOMIALS OVER A FIELD

Theorem:

Let F be a field and f (x) and g (x) be any two polynomial in F [x], not both of which are zero. Then f (x) and g (x) have a greatest common divisor d (x) which can be expressed in the form

$$d(x) = m(x) f(x) + n(x) g(x)$$

for polynomials m (x) and n (x) in F [x].

Proof:

Consider the set

$S = \{s(x) f(x) + t(x) g(x): s(x), t(x) \in F[x]\}$. ...(1)

We claim that S is an ideal of F [x]. The proof is as follows.

Let $s_1(x) f(x) + t_1(x) g(x)$ and $s_2(x) f(x) + t_2(x) + g(x)$ be any two elements of S.

Then $[s_1(x) f(x) + t_1(x) g(x)] - [s_2(x) f(x) + t_2(x) g(x)]$

$= [s_1(x) - s_2(x)] f(x) + [t_1(x) - t_2(x)] g(x) \in S$

since $s_1(x) - s_2(x)$ and $t_1(x) - t_2(x)$ are both members of F [x].

Also if a (x) be any member of F [x], then

$$\alpha(x) [s_1(x) f(x) + t_1(x) g(x)]$$
$$= [a(x) s_1(x)] f(x) + [a(x) t_1(x)] g(x) \in S.$$

Therefore S is an ideal of F [x]. Now every ideal in F [x] is a principal ideal. Therefore there exists an element d (x) in S such that every element in S is a multiple of d (x).

Since d (x) $\in$ S, therefore from (1) we see that there exist elements m (x), n (x) $\in$ F [x] such that

$$d(x) = m(x) f(x) + n(x) g(x).$$

Now F [x] is a ring with unity element 1.

∴ Putting $s(x) = 1$, $t(x) = 0$ in (1), we see that $f(x) \in S$. Also putting $s(x) = 0$, $t(x) = 1$ in (1), we see that $g(x) \in S$.

Now f (x), g (x) are elements of S. Therefore they are both multiples of d (x). Hence $d(x) \mid f(x)$ and $d(x) \mid g(x)$.

Now suppose $c(x) \mid f(x)$ and $c(x) \mid g(x)$.

Then $c(x) \mid [m(x) f(x)]$ and $c(x) \mid [n(x) g(x)]$. Therefore c (x) is also a divisor of $m(x) f(x) + n(x) g(x)$ *i.e.*, c (x) is a divisor of d (x).

Thus d (x) is a greatest common divisor of f (x) and g (x).

Note: If d (x) is a greatest common divisor of f (x) and g (x) then any associate of d (x) *i.e.*, k d (x) where $\neq k \in F$ will also be a greatest common divisor of f (x) and g (x). In particular if $0 \neq b$ is the leading coefficient of the polynomial d (x), then the monic polynomial $b^{-1} d(x)$ will also be a greatest common divisor of f (x) and g (x). *Often while defining greatest common divisor of two polynomials over a field we include one more condition in our definition that the greatest common divisor should be a monic polynomial.* The advantage of this extra condition is that now we shall get a unique greatest common divisor as shown below:

Supposed $d_1(x)$ and $d_2(x)$ are two monic polynomials and each is a greatest common divisor of f (x) and g (x). Then $d_1(x) \mid d_2(x)$ and $d_2(x) \mid d_1(x)$. Therefore $d_1(x_1)$ and $d_2(x)$ are associates and we have $d_1(x) = ud_2(x)$ for some $0 \neq u \in F$. Since $d_1(x)$ and $d_2(x)$ are both monic, therefore, $u = 1$.

3.4 UNIQUE FACTORIZATION DOMAIN

Definition: *An integral domain, R, with unity element 1 is a unique factorization domain if*

(a) any non-zero element in R is either a unit or can be written as the product of a finite number of irreducible (prime) elements of R;

(b) the decomposition in part (a) is unique upto the order and associates of the irreducible elements.

Thus, if R is a unique factorization domain and if $a \neq 0$ is a non-unit in R, then a can be expressed as a product of a finite of prime elements of R. Also if

$$a = p_1 p_2 p_3 \cdots p_n = p_1' p_2' p_3' \cdots p_m'$$

where the p_i and p_j' are prime elements of R, then m = n and each p_i, $1 \leq i \leq n$ is an associate of some p_j' $1 \leq j \leq m$ and conversely each p_k' is an associate of some p_l.

3.5 THE UNIQUE FACTORIZATION THEOREM FOR POLYNOMIALS OVER A FIELD

We shall now prove that every polynomial over a field can be factored uniquely into irreducible factors. Before stating the main factorization theorem, we shall give two preliminary theorems that are needed for its proof.

Theorem 1:

Let f (x), g (x) and h (x) be polynomials in F [x] for a field F. If f (x) | g (x) h (x) and the greatest common divisor of f (x) and g (x) is 1, then f (x) | h (x).

Proof:

If the greatest common divisor of f (x) and g (x) is 1, then by theorem of there exist polynomials m (x) and n (x) ∈ F [x] such that 1 = m(x) f(x) + n (x) g (x). Multiplying both members of this equations by h (x), we get

$$h(x) = m(x)\, f(x)\, h(x) + n(x)\, g(x)\, h(x) \qquad ...(1)$$

But f (x) | g (x) h (x), so there exists a polynomial q (x) ∈ F [x] such that g (x) h (x) = q (x) f (x).

Substituting this value of g (x) h (x) in (1), we get

$$h(x) = m(x)\, f(x)\, h(x) + n(x)\, q(x)\, f(x)$$

$$= f(x)\,[m(x)\, h(x) + n(x)\, q(x)],$$

which shows that f (x) is a divisor of h (x).

Hence the theorem.

Theorem 2:

If f (x) is an irreducible polynomial in F [x] for a field F and f (x) | g (x) h (x) where g (x), h (x) ∈ F [x] then f (x) divides at least one of g (x) or h (x).

Proof:

Suppose that f (x) does not divide g (x). Since f (x) is prime therefore f (x) does not divide g (x) implies that f (x) and g (x) are relatively prime. Therefore the greatest common divisor of f (x) and g (x) is 1. Hence by theorem 1, we get that f (x) | h (x).

Corollary: *If f (x) an irreducible polynomial in F [x] for a field F, and if f (x) divides the product $g_1(x)\,g_2(x)\ldots g_n(x)$ of polynomials in F [x], then f (x) divides $g_i(x)$ for some i, $1 \le i \le n$.*

This result follows immediately by repeated application of theorem.

The Unique Factorization Theorem for Polynomials Over a Field

Theorem:

Let f (x) be a non-zero polynomial in F [x], where F is a field. Then either f (x) is a unit n F [x] or $f(x) = a(x)\,p_1(x)\,p_2(x)\ldots p_m(x)$, where each $p_i(x)$, $1 \le i \le m$, is an irreducible monic polynomial in F [x] and $a \in F$ is the leading coefficient of f (x). Further the factors $p_1(x), p_2(x), \ldots, p_m(x)$ are unique except for the order in which they appear.

Proof:

We shall prove the theorem in two parts. First we shall prove that f (x) can be factored as required, and then we shall show that the factors are unique.

Let f (x) be a non-zero element of F [x]. Then either f (x) is a unit in F [x] *i.e.*, deg f (x) is 0 or deg f (x) > 0. If deg f (x) > 0, and the leading coefficient of f (x) is a we are to prove that f (x) can be expressed as a product of a and a finite number of irreducible monic polynomials in F [x]. The proof will be any induction on the degree of f (x).

Suppose f (x) is of degree one. Let f (x) = b + ax for a, b $\in$ F and a $\neq$ 0. We can write $f(x) = a(a^{-1}b + x)$. Therefore the theorem holds in the case where f (x) has degree one since $a^{-1}b + x$ is irreducible and monic.

Now assume, as the induction hypothesis, that every polynomial of degree less than n can be factored as stated in the theorem. Consider an arbitrary polynomial f (x) of degree n having a as its leading coefficient. We can write $f(x) = af_1(x)$, where $f_1(x) = a^{-1} f(x)$ and $f_1(x)$ is monic. If f (x) is irreducible, then $f_1(x)$ is also irreducible and the theorem holds. If f (x) is reducible, then it can be factored as f (x) = g (x) h (x) where neither g (x) nor h (x) is a unit in F [x]. Now the degree of f (x) is equal to the sum of the degrees of g (x) and h (x). Also g (x) and h (x) are not units in F [x], so each of them must be of degree one or larger. Hence both g (x) and h (x) have degrees less than n. Therefore by our induction hypothesis we can write

$$g(x) = c\alpha_1(x)\,\alpha_2(x)\ldots\alpha_s(x),\quad h(x) = d\,\beta_1(x)\,\beta_2(x)\ldots\beta_t(x)$$

where each $\alpha_i(x)$ and each $\beta_j(x)$ is monic and irreductivle and where c and d are leading coefficients of g (x) and h (x) respectively.

Thus $f(x) = cd\alpha_1(x)\,\alpha_2(x)\,\alpha_3(x)\ldots\alpha_s(x)\,\beta_1(x)\,\beta_2(x)\ldots\beta_t(x)$.

Since the leading coefficient of f (x) is a, therefore we must have a = cd because each $\alpha(x)$ and each $\beta(x)$ is monic. Therefore

$$f(x) = a\,\alpha_1(x)\,\alpha_2(x)\ldots\alpha_s(x)\,\beta_1(x)\,\beta_2(x)\ldots\beta_i(x).$$

The factorization of f (x) satisfies the requirements of the theorem. Hence the theorem holds for all polynomials of degree n, and by the principle of induction, for all polynomials of arbitrary degree.

In order to prove that the factors are unique, let us suppose that $f(x) = a\,p_1(x)\,p_2(x)\ldots p_m(x) = a\,q_1(x)\,q_2(x)\ldots q_n(x)$ where each p (x) and each q (x) is irreducible and monic. Then we shall prove that n = m and each p (x) is equal to some q (x) and each q (x) is equal to some p (x). From these two decompositions of f (x), we have

$$p_1(x)\,p_2(x)\ldots p_m(x) = q_1(x)\,q_2(x)\ldots q_n(x).$$

Now $\quad p_1(x) \mid p_1(x)\,p_2(x)\ldots p_m(x)$. Therefore

$$p_1(x) \mid q_1(x)\,q_2(x)\ldots q_n(x).$$

By Corollary to theorem of this article $p_1(x)$ must divide at least one of $q_1(x), q_2(x), \ldots, q_n(x)$. Since F [x] is a commutative ring, therefore without loss of generality we may suppose that $p_1(x)$ divides $q_1(x)$. But $p_1(x)$ and $q_1(x)$ are both irreducible polynomials in F [x] and $p_1(x) \mid q_1(x)$. Therefore $p_1(x)$ and $q_1(x)$ must be associates and we have $q_1(x) = up_1(x)$ where u is a unit in F [x] *i.e.*, u is a non-zero element of F. Since $q_1(x)$ and $p_1(x)$ are monic therefore u must be equal to 1 and we have $p_1(x) = q_1(x)$. Thus we have

$$p_1(x)\,p_2(x)\ldots p_m(x) = p_1(x)\,q_2(x)\ldots q_n(x).$$

Cancelling $0 \,' \, p_1(x)$ from both sides, we get

$$p_2(x)\,p_3(x)\ldots p_m(x) = q_2(x)\,q_3(x)\ldots q_n(x) \qquad \ldots(1)$$

Now we can repeat the above argument on the relation (1) with $p_2(x)$. If n > m, then after m steps the left hand side becomes 1 while the right hand side reduces to a product of a certain number of q (x) (the excess of n over m). But the q (x) are irreducible polynomials so they are not units of F [x] *i.e.*, they are not polynomials of zero degree.

So their product will be a polynomial of degree ≥ 1. So it cannot be equal to 1. Therefore n cannot be greater than m. Then $n \leq m$. Similarly interchanging the roles of p (x) and q (x), we get $m \leq n$. Hence m = n.

Also in the above process we have shown that every p (x) is equal to some q (x) and conversely every q (x) is equal to some p (x). Hence the theorem has been completely established.

Thus, we can say that *the ring of polynomials over a field is a unique factorization domain.*

3.6 VALUE OF A POLYNOMIAL AT X = C

Definition: *Let* $f(x) = a_0 + a_1x + a_2x^2 + \ldots + a_nx^n$ *be a polynomial in* $F[x]$ *for an arbitrary field F and let c be an element of F. Then* $f(c) = a_0 + a_1c + a_2c^2 + \ldots + a_nc^n$, *where the indicated addition and multiplication are the operations in F, is called the value of* $f(x)$ *at* $x = 0$. Obviously f (c) is an element of F.

Zeros of a Polynomial

Definition: *If* $f(x)$ *is a polynomial in* $F[x]$ *for an arbitrary field F, and* $f(c) = 0$ *for an element* $c \in F$, *then c is called a zero of* $f(x)$.

Polynomial Equations and their Roots

Definition: *Let* $f(x)$ *be a polynomial of degree n over a field F. We say that* $f(x) = 0$ *is an equation over the field F and n is the degree of the equation.*

If c is a zero of the polynomial $f(x)$, *then c is a root of the equation* $f(x) = 0$. A root of an equation is also called a *solution* of the equation.

Remainder Theorem:

If $f(x) \in F[x]$ *and* $a \in F$, *for any field F, then* $f(a)$ *is the remainder when* $f(x)$ *is divided by* $(x - a)$.

Proof:

By division algorithm there exist polynomials q (x) and r (x) such that $f(x) = q(x)(x - a) + r(x)$, where either $r(x) = 0$ or deg r (x) is less than the degree of $x - a$. But the degree of $(x - a)$ is 1. Therefore r (x) has degree 0 or no degree. Hence r (x) is a constant polynomial *i.e.*, r (x) is simply an element, say, r in F. Thus $f(x) = q(x)(x - a) + r$. Putting $x = a$ in this relation, we get $f(a) = q(a)(a - a) + r \Rightarrow f(a) = r$.

Cor. Factor Theorem:

If $f(x) \in F[x]$ *and* $a \in F$, *for a field F, then* $x - a$ *divides* $f(x)$ *if and only if* $f(a) = 0$.

Proof:

By remainder theorem, f (a) is the remainder when f (x) is divided by $(x - a)$. Therefore if $f(a) = 0$, then $(x - a)$ divides f (x).

Conversely, if f (x) is divisible by (x – a) we get

$$f(x) = (x - a)\, q(x).$$

Putting x = a, we get f (a) = (a – a) q (a) = 0 q (a) = 0.

Example 1:

Show that the polynomial $x^3 - 9$ is reducible over the ring of integers modulo 11.

Solution:

Let R be the ring of integers modulo 11.

Then R is the ring

$$(\{0, 1, 2, ..., 10\}, +_{11}, \times_{11}).$$

Let $\quad f(x) = x^3 - 9.$

In the ring of integers modulo 11, we have

$$2 +_{11} 9 = 0.$$

$\therefore \quad -9 = 2.$

So over the ring of integers modulo 11, we have

$$f(x) = x^3 - 9 = x^3 + 2.$$

In order to show that f (x) is reducible over the ring of integers modulo 11, we shall produce a proper factor of f (x) over the ring of integers modulo 11.

We have

$$f(4) = 4^3 +_{11} 2 = (4 \times_{11} 4 \times_{11} 4) +_{11} 2 = 9 +_{11} 2 = 0.$$

So by factor theorem x – 4 *i.e.*, x + 7 is a factor of f (x) over the ring of integer modulo 11. But x + 7 is not a polynomial of zero degree over the ring of integers modulo 11 and so x + 7 is not a unit of the polynomial ring R [x]. Consequently x + 7 is a proper factor of f (x) over the ring of integers modulo 11.

Hence f (x) = $x^3 - 9 = x^3 + 2$ is reducible over the ring of integers modulo 11.

Example 2:

Resolve $x^2 + 1$ into factors over the field Z_5.

Solution:

The field Z_5 is $(\{0, 1, 2, 3, 4\}, +_5, \times_5)$.

Let $\quad f(x) = x^2 + 1.$

We have $f(0) = 0^2 +_5 1 = 1 \neq 0,$

$f(1) = 1^2 +_5 1 = 2 \neq 0$

$f(2) = 2^2 +_5 1 = 4 +_5 1 = 0,$

$f(3) = 3^2 +_5 1 = (3 \times_5 3) +_5 1 = 4 +_5 1 = 0,$

$f(4) = 4^2 +_5 1 = (4 \times_5 4) +_5 1 = 1 +_5 1 = 2 \neq 0.$

By factor theorem, we know that if f (x) is a polynomial over the field F and $a \in F$, then x – a divides f (x) if and only if f (a) = 0.

So the only factors of $f(x) = x^2 + 1$ over field Z_5 are x – 2 and x–3.

Here – 2 and – 3 are the additive inverse of 2 and 3 respectively in the field Z_5. In the field Z_5, we have

$$2 +_5 3 = 0.$$

$\therefore -2 = 3$ and $-3 = 2$.

So over the field Z_5, we have

$$x - 2 = x + 3 \text{ and } x - 3 = x + 2.$$

Thus over the field Z_5, we have

$$x^2 + 1 = (x - 2)(x - 3) = (x + 3)(x + 2).$$

Over the field Z_5,

$$x^2 + 1 = (x + 2)(x + 3)$$

Example 3:

Find the solution of the equation 3x = 2 in the field $(Z_7, +_7, \times_7)$.

Solution:

The field $(Z_7, +_7, \times_7)$ is the field $(\{0, 1, 2, 3, 4, 5, 6\}, +_7, \times_7)$.

If $x \in Z_7$ and x = 3, we have

$$3x = x +_7 x +_7 x = 3 +_7 3 +_7 3 = 2.$$

Thus x = 3 is a solution of the equation 3x = 2 in the field $(Z_7, +_7, \times_7)$.

Also no other element of field Z_7 satisfies the equation 3x = 2.

If x = 0,	we have	3x = 0,
if x = 1,	we have	3x = 1,
if x = 2,	we have	3x = 6,
if x = 4,	we have	3x = 5,

if $x = 5$, we have $3x = 1$,

and if $x = 6$ we have $3x = 4$.

Hence $x = 3$ is the only solution of the equation $3x = 2$ in the given field $(Z_7, +_7, \times_7)$.

Example 4:

If $f(x) = 3x^7 + 2x + 3$, $g(x) = 5x^3 + 2x + 6$ be two polynomials over the field $Z_7 = (\{0, 1, 2, 3, 4, 5, 6\}, +_7, \times_7)$, determine

(i) $\frac{d}{dx} f(x)$ (ii) $f(x) \cdot g(x)$, and (iii) $f(x) + g(x)$.

Solution:

(i) We have

$= (7 +_7 7 +_7 7)\, x^6 + (1 +_7 1)$

(ii) We have $f(x)\, g(x)$

$= (3 + 2x + 3x^7)(6 + 2x + 5x^3)$

$= (3 \times_7 6) + [(3 \times_7 2) +_7 (2 \times_7 6)]\, x + (2 \times_7 2)\, x^2$

$+ (3 \times_7 6)\ (2 \times_7 5)\, x^4 + (3 \times_7 6)\, x^7 + (3 \times_7 2)\, x^8 + (3 \times_7 5)\, x^{10}$

$= 4 + 4x + 4x^4 + 3x^4 + 4x^7 + 6x^8 + x^{10}$

(iii) We have $f(x) + g(x)$

$= (3 + 2x + 3x^7) + (6 + 2x + 5x^3)$

$= (3 +_7 6) + (2 +_7 2)\, x + (0 +_7 5)\, x^3 + (3 +_7 0)\, x^7$

$= 2 + 4x + 5x^3 + 3x^7$

$= 3x^7 + 5x^3 + 4x + 2.$

Example 5:

Let $f(x) = x^6 + 3x^3 + 4x^2 - 3x + 2$ and $g(x) = x^2 + 2x - 3$ be in $Z_7[x]$. Find

(i) Sum and product of $f(x)$ and $g(x)$ in $Z_7[x]$.

(ii) Two polynomials $q(x)$ and $r(x)$ in $Z_7[x]$ such that

$f(x) = q(x) + r$ with degree $r(x) < 2$.

Solution:

The field Z_7 is $(\{0, 1, 2, 3, 4, 5, 6), +_7, \times_7)$.

In Z_7, $3 +_7 4 = 0$ so that $4 = -3$.

$\therefore$ we can write

$$f(x) = x^6 + 3x^5 + 4x^2 - 3x + 2 = x^6 + 3x^5 + 4x^2 + 4x + 2$$

and $$g(x) = x^2 + 2x - 3 = x^2 + 2x + 4.$$

(i) We have $f(x) + g(x)$

$$= (2 + 4x + 4x^2 + 3x^5 + x^6) + (4 + 2x + x^2)$$
$$= (2 +_7 4) + (4 +_7 2)\,x + (4 +_7 1)\,x^2 + 3x^5 + x^6$$
$$= 6 + 6x + 5x^2 + 3x^5 + x^6$$
$$= x^6 + 3x^5 + 5x^2 + 6x + 6.$$

Also $f(x)\,g(x)$

$$= (2 + 4x + 4x^2 + 3x^5 + x^6)\,(4 + 2x + x^2)$$
$$= (2 \times_7 4) + [(2 \times_7 2) +_7 (4 \times_7 4)]\,x + [(2 \times_7 1) +_7 (4 \times_7 2)$$
$$+_7 (4 \times_7 4)]\,x^2 + [(4 \times_7 1) +_7 (4 \times_7 2)]\,x^3 + (4 \times_7 1)\,x^4$$
$$+ (3 \times_7 4)\,x^5 + [(3 \times_7 2) +_7 (1 \times_7 4)]\,x^6$$
$$+ [(3 \times_7 1) +_7 (1 \times_7 2)]\,x^7 + (1 \times_7 1)\,x^8$$
$$= 1 + 6x + 5x^2 + 5x^3 + 4x^4 + 5x^5 + 3x^6 + 5x^7 + x^8$$
$$= x^8 + 5x^7 + 3x^6 + 5x^5 + 4x^4 + 5x^3 + 5x^2 + 6x + 1.$$

In $Z_7\,[x]$, let us divide $f(x)$ by $g(x)$ by long division method.

$$\begin{array}{r|l}
 & x^4 + x^3 + x^2 + x + 5 \\ \hline
x^2 + 2x + 4 & x^6 + 3x^5 + 4x^2 + 4x + 2 \\
 & x^6 + 2x^5 + 4x^4 \\ \hline
 & x^5 + 3x^4 + 4x^2 + 4x + 2 \\
 & x^5 + 2x^4 + 4x^3 \\ \hline
 & x^4 + 3x^3 + 4x^2 + 4x + 2 \\
 & x^4 + 2x^3 + 4x^2 \\ \hline
 & x^3 + 4x + 2 \\
 & x^3 + 2x^2 + 4x \\ \hline
 & 5x^2 + 2 \\
 & 5x^2 + 3x + 6 \\ \hline
 & 4x + 3
\end{array}$$

$[\because$ in Z_7, $-4 = 3]$

$[\because$ in Z_7, $-3 = 4$ and $2 - 6 = 2 +_7 (-6) = 2 +_7 1 = 3]$

The degree of the remainder $4x + 3$ is 1 which is less than 2 *i.e.*, the degree of the divisor $g(x) = x^2 + 2x + 4$.

Thus we have

$x^6 + 3x^5 + 4x^2 + 4x + 2 = (x^4 + x^3 + x^2 + x + 5)(x^2 + 2x + 4) + 4x + 3$

i.e., $f(x) = q(x)\, g(x) + r(x).$

Example 6:

Show that $f(x) = x^2 + 8x - 2$ is irreducible over the field of rational numbers Q. Is irreducible over real? Give reasons for your answer.

Solution:

First we solve the equation $f(x) = 0$ over the field C of complex numbers.

We have $x^2 + 8x - 2 = 0$

$$\Rightarrow \quad x = \frac{-8 \pm \sqrt{(64+8)}}{2} = -4 \pm 3/2.$$

Thus, the only proper factors of $f(x) = x^2 + 8x - 2$ over the field of complex numbers are

$$x - (-4 + 3\sqrt{2}) \text{ and } x - (-4 - 3\sqrt{2}).$$

Since neither $-4 + 3\sqrt{2}$ nor $-4 - 3\sqrt{2}$ is a rational number, therefore neither $x (-4 + 3\sqrt{2})$ nor $x - (-4 - 3\sqrt{2})$ is a polynomial over the field Q. Hence f (x) has no proper factor over the field of rational numbers Q and so f (x) is irreducible over the field Q.

But both $-4 + 3\sqrt{2}$ and $-4 - 3\sqrt{2}$ are real numbers and so both $x - (-4 + 3\sqrt{2})$ and $x - (-4 - 3\sqrt{2})$ are polynomials over the field of real numbers R. Thus $x - (-4 + 3\sqrt{2})$ and $x (-4 - 3\sqrt{2})$ are proper factors of $f(x) = x^2 + 8x - 2$ over the field of real numbers R. Hence $f(x) = x^2 + 8x - 2$ is reducible over the field R.

So $f(x) = x^2 + 8x - 2$ is not irreducible over the field of real number R.

Example 7:

Let $f(x) = 2x^4 + 3x^3 + 2$ and $g(x) = 3x^5 + 4x^3 + 2x^3 + 3$ be two polynomials over the field $Z_5 = (\{0, 1, 2, 3, 4)\}, +_5, \times_5)$.

Determine (i) $\frac{d}{dx} f(x)$, (ii) $f(x).\, g(x)$.

Solution:

(i) We have

$$\frac{d}{dx} f(x) = 4\,(2)\,x^3 + 3\,(3)\,x^2$$

$$= (2 +_5 2 +_5 2 +_5 2)\,x^3 + (3 +_5 3 +_5 3)\,x^2 = 3x^3 + 4x^2.$$

(ii) We have f (x) g (x)

$$= (2 + 3x^3 + 2x^4)(3 + 2x^2 + 4x^3 + 3x^5)$$

$$= (2 \times_5 3) + (2 \times_5 2)\,x + [(2 \times_5 4) +_5 (3 \times_5 3)]\,x^3$$

$$+ (2 \times_5 3)\,x^4 + [(2 \times_5 3) +_5 (3 \times_5 2)]\,x^5$$

$$+ [(3 \times_5 4) +_5 (2 \times_5 2)]\,x^6 + (2 \times_5 4)\,x^7 + (3 \times_5 3)\,x^8 + (2 \times_5 3)\,x^9$$

$$= 1 + 4x^2 + 2x^3 + x^4 + x^4 + 2x^5 + x^6 + 3x^7 + 4x^8 + x^9.$$

where the quotient $q(x) = x^4 + x^3 + x^2 + x + 5$.

and the remainder $r(x) = 4x + 3$.

Example 8:

Show that the polynomial $x^2 + x + 4$ is irreducible over F, the field of integers modulo 11.

Solution:

The field F is $\{0, 1, ..., 10\}, +_{11}, \times_{11})$.

Let $f(x) = x^2 + x + 4$.

If $a \in F$, then by a^n we shall mean $a \times_{11} a \times_{11} a \times_{11} a$... upon n times.

Now $f(0) = 0^2 +_{11} 0 +_{11} 4 = 4$, $f(1) = 1^2 +_{11} 1 +_{11} 4 = 6$.

$f(2) = 2^2 +_{11} 2 +_{11} 4 = 10$, $f(3) = 3^2 +_{11} 3 +_{11} 4 = 5$, $f(4) = 2$,

$f(5) = 1$, $f(6) = 6^2 +_{11} 6 +_{11} 4 = 2$, $f(7) = 5$, $f(8) = 10$, $f(9) = 6$, $f(10) = 4$.

Since $f(a) \neq 0 \ \forall \ a \in F$, therefore by factor theorem $x - a$ does not divide $f(x) \ \forall \ a \in F$. Therefore f (x) has no proper divisors in F [x]. Hence f (x) is irreducible over F.

Example 9:

Resolve $x^4 + 4$ into factors over the field

$(\{0, 1, 2, 3, 4\}, +_5, \times_5)$.

Solution:

Let F be the field $(\{0, 1, 2, 3, 4\}, +_5, \times_5\}$.

Let $\quad f(x) = x^4 + 4 \in F[x]$.

We have

$f(0) = 0^4 +_5 4 = 4 \neq 0,\ f(1) = 1^4 +_5 4 = 1 +_5 4 = 0,$

$f(2) = 2^4 +_5 4 = (2 \times_5 2 \times_5 2 \times_5 2) +_5 4 = 1 +_5 4 = 0,$

$f(3) = 3^4 +_5 4 = 1 +_5 4 = 0$ and $f(4) = 4^4 +_5 4 = 1 +_5 4 = 0.$

By factor theorem, we know that if f (x) is a polynomial over the field F and a ∈ F, then x – a is a factor of f (x) iff f (a) = 0.

So the only factors of f (x) over the given field F are x – 1, x – 2, x – 3 and x – 4.

Now in the given field $(\{0, 1, 2, 3, 4), +_5, \times_5)$, we have

$$1 +_5 4 = 0 \text{ and } 2 +_5 3 = 0.$$

∴ $-1 = 4$ and $-4 = 1$. Also $-2 = 3$ and $-3 = 2$.

Thus over the given field F, we have

$$x - 1 = x + 4,\ x - 2 = x + 3,\ x - 3 = x + 2 \text{ and } x - 4 = x + 1.$$

Hence over the given field $(\{0, 1, 2, 3, 4), +_5, \times_5)$, we have

$$x^4 + 4 = (x + 1)(x + 2)(x + 3)(x + 4).$$

Example 10:

If p is a prime integer, show that it need not be prime Gaussian integer.

Solution:

We shall give an example of prime integer which is not a prime Gaussian integer.

A complex number a + ib is called a Gaussian integer if its real and imaginary parts a and b are both in integers.

Consider the integer 5. In the ring of integers the only divisors of 5 are ± 1 and ± 5. Thus 5 has no proper divisors in the ring of integers and so 5 is a prime integer.

In the ring of Gaussian integers we can factorize 5 as

$$5 = (2 + i)(2 - i).$$

The only units of the ring of Gaussian integers are ± 1 and ± i. So neither 2 + i nor 2 – i is a unit of the ring of Gaussian integers.

Consequently 2 + i and 2 – i are proper divisors of 5 in the ring of Gaussian integers. Hence 5 is not a prime element in the ring of Gaussian integers.

Thus 5 is a prime integer, but it is not a prime Gaussian integer.

3.7 QUOTIENT RINGS OR RINGS OF RESIDUE CLASSES

Suppose R is an arbitrary ring and S is an ideal (two sided ideal) in R. Then S is a subgroup of the additive abelian group of R. We can form the cosets (right as well as left) of S in R. Since R is an abelian additive group, therefore if $a \in R$, then the right coset $S + a$ will be equal to the corresponding left coset $a + s$. Thus, we shall call $S + a$ as simply a coset of S in R. We remember from our study of cosets in group theory, that if $a, b \in R$, Then

$$S + a = S + b \Rightarrow a - b \in S.$$

The cosets of S in R are called the residue classes of S in R. We denote the set of all residue classes of S in R by the symbol R/S.

Thus $\quad R/S = \{S + a : a \in R\}$.

We shall now impose a ring structure on the set R/S by defining addition and multiplication of residue classes.

Theorem:

If S is an ideal of a ring R, then the set

$$R/S = \{S + a : a \in R\}$$

of all residue classes of S in R forms a ring for the two compositions in R/S defined as follows:

$(S + a) + (S + b) = S + (a + b)$ [*Addition of residue classes*]

$(S + a)(S + b) = S + ab.$ [*Multiplication of residue classes*]

Proof:

Since $S + (a + b)$ and $S + ab$ are also residue classes of S in R, therefore R/S is closed with respect to addition and multiplication of residue classes. First of all, we shall show that both addition and multiplication in R/S are well defined. For this we are to show that if $S + a = S + a'$ and $S + b = S + b'$, then

$$(S + a) + (S + b) = (S + a') + (S + b')$$

and $\quad (S + a)(S + b) = (S + a')(S + b')$.

We have $\quad S + a = S + a' \Rightarrow a' \in S + a$

and $\quad S + b = S + b' \Rightarrow b' \in S + b$.

Therefore there exist $\alpha, \beta \in S$ such that $a' = \alpha + a$, $b' = \beta + b$.

Now $\quad a' + b' = (\alpha + a) + (\beta + b) = (a + b) + (\alpha + \beta)$.

$\therefore \quad (a' + b') - (a + b) = \alpha + \beta \in S$.

$\therefore \quad S + (a' + b) + (S + b') = (S + a) + (S + b)$.

Thus addition in R/S is well defined.

Again $a'b' = (\alpha + a)(\beta + b) = \alpha b + \alpha\beta + a\beta + ab$

$= ab + \alpha\beta + \alpha b + a\beta.$

$\therefore$ $a'b' - ab = \alpha\beta + \alpha\beta + \alpha\beta \in S$. [Since S is an ideal therefore $\alpha, \beta \in S$ and $a, b \in R \Rightarrow \alpha b \in S, \alpha\beta \in S, \alpha\beta \in S$ and finally $\alpha\beta + \alpha\beta + a\beta \in S$].

Now since $a'b' - ab \in S$, therefore $S + a'b' = S + ab$

$\Rightarrow \quad (S + a')(S + b') = (S + a)(S + b).$

Hence multiplication in R/S is also well defined.

Associativity of addition in R/S. We have

$(S + a) + [(S + b) + (S + c)] = (S + a) + [S (b + c)]$

$= S + [a + (b + c)] = S + [(a + b) + c] = [S + (a + b)] + (S + c)$

$= [(S + a) + (S + b)] + (S + c).$

Commutativity of addition in R/S. We have

$(S + a) + (S + b) = S + (a + b) = S + (b + a) = (S + b) + (S + a).$

Existence of additive identity. We have $S = S + 0 \in R/S$.

If $S + a \in R/S$, then $(S + 0) + (S + a) = S + (0 + a) = S + a.$

$\therefore$ S is the additive identity.

Existence of additive inverse. Let $S + a \in R/S$.

Then $S + (-a) \in R/S$. Also we have

$[S + (-a)] + [S + a] = S + [(-a) + a] = S + 0 = S.$

$\therefore$ $S + (-a)$ or $S - a$ is the additive inverse of $S + a$.

Associativity of multiplication. We have

$[(S + a)(S + b)](S + c) = (S + ab)(S + c) = S + (ab)c$

$= S + a(bc) = (S + a)(S + bc) = (S + a)[(S + b)(S + c)].$

Distributivity of multiplication with respect to addition. We have

$(S + a)[(S + b) + (S + c)] = (S + a)[S + (b + c)]$

$= S + a(b + c) + S + (ab + ac) = (S + ab) + (S + ac)$

$= (S + a)(S + b) + (S + a)(S + c).$

Similarly, we can prove that

$[(S + b) + (S + c)](S + a) = (S + b)(S + a) + (S + c)(S + a).$

Hence R/S is a ring with respect to the two compositions. The residue class S + 0 or S is the zero element of this ring.

Note: The students should not confuse that by the multiplication (S + a) (S + b) of residue classes we mean the totality of elements obtained on multiplication of residue classes is a new composition which we have defined in the set R/S.

However, the addition (S + a) + (S + b) of residue classes as defined by us coincides with the totality of elements obtained on adding elements of S + a to the elements of S + b as can be easily seen:

$$(S + a) + (S + b) = S + (a + S) + b = S + (S + a) + b$$
$$= (S + S) + (a + b) = S + (a + b).$$

Example 1:

If R/S is a ring of residue classes of S in R, prove that

(i) If R is commutative, so also is R/S.

(ii) if R has a unity element 1 so also has R/S, namely S + 1.

Solution:

(i) Suppose R is a commutative ring Let S + a, S + b be any two elements of R/S. Then a, b ∈ R and ab = ba.

We have (S + a) (S + b) = S + ab = S + ba = (S + b) (S + a).

(ii) Suppose R is a ring with unit element 1. Then S + 1 ∈ R/S. If S + a is any element of R/S, we have

$$(S + 1)(S + a) = S + (1a) = S + a$$

and $$(S + a)(S + 1) = S + (a1) = S + a.$$

∴ S + 1 is the unit element of R/S

3.8 HOMOMORPHISM OF RINGS

Definition: *Homomorphism into: A mapping f from a ring R into a ring R' is said to be a homomorphism of R into R' if*

(i) f (a + b) = f (a) f (a) + f (b) ∀ a, b ∈ R

(ii) f (ab) = f (a) f (b) for all a, b, ∈ R.

Homomorphism onto: *A mapping f from a ring R onto a ring R' is said to be a homomorphism of R onto R' if*

(i) f (a + b) = f (b) ∀ a, b ∈ R.

(ii) f (ab) = f (a) f (b) $\forall$ a, b, $\in$ R.

Also then R' is said to be a homomorphic image of R.

Theorem 1:

If f is a homomorphism of a ring R into a ring R', then

(i) f (0) = 0', *where 0 is the zero element of the ring R and 0' is the zero element of R'.*

(ii) f (– a) = – f (a) $\forall$ a $\in$ R.

Proof:

(i) Let a $\in$ R. Then f (a) $\in$ R'. We have

f (a) + 0' = f (a) [$\therefore$ 0' is the additive identity of R']

= f (a + 0) = f (a) + f (0).

Now R' is a group with respect to addition. Therefore

f (a) + 0' = f (a) + f (0)

$\Rightarrow$ 0' = f (0). [by left cancellation law]

(ii) Let a be any element of R. Then – a $\in$ R.

We have 0' = f (0) = f [a + (– a)] = f (a) + f (– a).

$\therefore$ f (– a) is the additive inverse of f (a) in the ring R'. Thus

f (– a) = – f (a).

Theorem 2:

Let ϕ be a homomorphic mapping of a ring R into a ring R'. Let S' be the homomorphic image of R in R'. Then S' is a subring of R'.

Proof:

Since S' is the image of R in R' under the mapping ϕ, therefore ϕ(R) = S $\subseteq$ R'.

Let a', b' be any two elements of S'. Since S' = ϕ (R), therefore there exist elements a, b, $\in$ R such that ϕ (a) = a' ϕ (b) = b'.

We have a' – b' = ϕ (a) – ϕ (b) = ϕ (a – b).

[$\therefore$ ϕ is a homomorphism]

Now a – b $\in$ R is such that a' – b' = ϕ (a – b). Therefore

a' – b' $\in$ S'.

Further a'b' = ϕ (a) ϕ (b) = ϕ (ab) $\in$ S', since ab $\in$ R.

Thus $a', b' \in S'$

$\Rightarrow$ $a' - b' \in S'$ and $a'b' \in S'$.

Therefore S' is a subring of R'.

3.9 KERNEL OF A RING HOMOMORPHISM

Definition: *If f is a homomorphism of a ring R into a ring R', then the set S of all those elements of R which are mapped onto the zero element of R' is called the kernel of the homomorphism f.*

Thus, if f is a homomorphism of R into R', then S is the kernel of f if $S = \{x \in R: f(x) = 0'$ where 0' is the zero element of R'$\}$.

Theorem 1:

If f is a homomorphism of a ring R into a ring R' with kernel S, then S is an ideal of R.

Proof:

Let f be a homomorphism of a ring R into a ring R'. Let 0, 0' be the zero elements of R, R' respectively. Let S be the kernel of f. Then $S = \{x \in R: f(x) = 0'\}$.

Since $f(0) = 0'$, therefore at least $0 \in S$. Thus S is not empty.

Let $a, b, \in S$. Then $f(a) = 0'$, $f(b) = 0'$.

We have $f(a - b) = f[a + (-b)] = f(a) + f(-b)$

$= f(a) - f(b) = 0' - 0' = 0'$.

$\therefore a - b \in S$.

Also if r be any element of R, then

$f(ar) = f(a) f(r) = 0' f(r) = 0'$

and $f(ra) = f(r) f(a) = f(r) 0' = 0'$.

$\therefore$ $ar \in S$, $ra \in S$.

Thus $a, b \in S$, $r \in R$

$\Rightarrow$ $(a - b) \in S$, $ar \in S$, $ra \in S$.

$\therefore$ S is an ideal of R.

Theorem 2:

The homomorphism of ϕ of a ring R into a ring R' is an isomorphism of R into R' if and only if $I(\phi) = (0)$, where $I(\phi)$ denotes the kernel of ϕ.

Proof:

Let ϕ be a homomorphism of a ring R into a ring R'. Let 0, 0' be the zero elements of R, R' respectively. Let S = I (ϕ) be kernel of ϕ. Then S is an ideal of R and

$$S = \{a \in R: \phi(a) = 0'\}.$$

Suppose ϕ is an isomorphism of a ring R into R'. Then ϕ is one-one. Let a $\in$ S. Then

$\phi(a) = 0'$ [by def. of kernel]

$\Rightarrow \phi(a) = \phi(0)$ [$\because \phi(0) = 0'$]

$\Rightarrow a = 0.$ [$\because \phi$ is one-one]

Thus $a \in S \Rightarrow a = 0$. In order words 0 is the only element of R which belongs to S. Therefore S = (0).

Conversely suppose that S = (0). Then to prove that ϕ is an isomorphism of R into *i.e.*, to prove that ϕ is one-one.

If a, b $\in$ R, then $\phi(a) = \phi(b)$

$\Rightarrow \quad \phi(a) - \phi(b) = 0'$ [$\because \phi(a), \phi(b)$ are in the ring R']

$\Rightarrow \quad \phi(a - b) = 0'$ [$\because \phi$ is a homomorphism]

$\Rightarrow \quad a - b \in S$ [by def. of kernel]

$\Rightarrow \quad a - b = 0$ [$\because S = (0)$]

$\Rightarrow \quad a = b.$

$\therefore \phi$ is one-one. Hence f is an isomorphism of R into R'.

Theorem 3:

Suppose R is a ring, S an ideal of R. Let f be a mapping from R to R/S defined by f (a) = S + a $\forall$ a $\in$ R. Then f is a homomorphism of R onto R/S.

Proof:

Consider the mapping f: R $\rightarrow$ R/S such that

$$f(a) = S + a \ \forall\ a \in R.$$

Let S + x be any element of R/S. Then x $\in$ R.

We have f (x) = S + x. Therefore the mapping is onto R/S.

Let a, b $\in$ R. Then

$$f(a + b) = S + (a + b) = (S + a) + (S + b) = f(a) + f(b),$$

Also $\quad f(ab) + S + ab = (S + a)(S + b) = f(a) f(b).$

$\therefore$ if is a homomorphism of R onto R/S.

Thus every quotient ring of is a homomorphic image of the ring.

Fundamental Theorem on Homomorphism of Rings

Theorem:

Every homomorphic image of a ring R is isomorphic to some residue class ring (quotient ring) thereof.

Proof:

Let R' be the homomorphic image of a ring R and f be the corresponding homomorphism. Then f is a homomorphism of R onto R'. Let S be the kernel of this homomorphism. Then S is an ideal of R. Therefore R/S is a ring of residue classes of R relative to S. We shall prove that $R/S \cong R'$.

If $a \subset R$, then $S + a \in R/S$ and $f(a) \in R'$. Consider the mapping $\phi: R/S \to R'$ such that

$$\phi(S + a) = f(a) \ \forall \ a \in R.$$

First we shall show that the mapping ϕ is well defined *i.e.*, if $a, b \in R$ and $S + a = S + b$, then $\phi(S + a) = \phi(S + b)$.

We have $S + a = S + b$

$\Rightarrow \quad a - b \in S$

$\Rightarrow \quad f(a - b) = 0'$ [*i.e.*, zero element of R']

$\Rightarrow \quad f[a + (-b)] = 0' \Rightarrow f(a) + f(-b) = 0'$

$\Rightarrow \quad f(a) - f(b) = 0'$

$\Rightarrow \quad f(a) = f(b) \Rightarrow \phi(S + a) = \phi(S + b).$

$\therefore$ ϕ is well-defined.

ϕ is one-one. We have $\phi(S + a) = \phi(S + b)$

$\Rightarrow \quad f(a) = f(b) \Rightarrow f(a) - f(b) = 0'$

$\Rightarrow \quad a - b \in S$ [$\because$ S is kernel of f]

$\Rightarrow \quad S + a = S + b.$

$\therefore$ ϕ is one-one.

ϕ is onto R'. Let y be any element of R'. Then $y = f(a)$ *for some* $a \in R$ because f is onto R'. Now $S + a \in R/S$ and we have

$\phi(S + a) = f(a) = y$. Therefore ϕ is onto R'.

Finally we have

$$\phi[(S + a) + (S + b) = \phi[S + (a + b)] = f(a + b)$$

$= f(a) + f(b) = \phi(S + a) + \phi(S + b).$

Also $\phi[(S + a)(S + b)] = \phi(S + ab) = f(ab) = f(a) f(b)$

$= [\phi(S + a)][\phi(S + b)].$

$\therefore$ ϕ is an isomorphism of R/S onto R'.

Hence $R/S \cong R'$.

Example 1:

If R is a ring with unit element 1 and ϕ is a homomorphism of R onto R' prove that $\phi(1)$ is the unit element of R'.

Solution:

Since f is a homomorphism of R onto R', therefore R' is a homomorphic image of R. If 1 is the unity element of R, then $\phi(1) \in R'$. Let a' be any element of R'. Then $a' = \phi(a)$ for some $a \in R$ since ϕ is onto R'. We have

$$\phi(1)\, a' = \phi(1)\, \phi(a) = \phi(1a) = \phi(a) = a'$$

and $$a'\, \phi(1) = \phi(a)\, \phi(1) = \phi(a1) = \phi(a) = a'.$$

$\therefore$ $\phi(1)$ is the unity element of R'.

Example 2:

If r is a ring with unit element 1 and ϕ is a homomorphism of R into an integral domain R' such that kernel of ϕ i.e., $I(\phi) \neq R$, then prove that $\phi(1)$ is the unit element of R'.

Solution:

ϕ is a homomorphism of a ring R into an integral domain R'. Then kernel of ϕ.

$$= I(\phi) = \{x : x \in R \text{ and } \phi(x) = 0 \in R'\}.$$

Since $I(\phi) \neq R$, therefore there exists an element $a \in R$ such that

$$\phi(a) \neq 0 \in R'.$$

We have $\phi(1)\, \phi(a) = \phi(1a) = \phi(a)$.

Now let b' be any element of R'. We have

$$\phi(a)\, b' = \phi(a)\, b'$$

$\Rightarrow$ $\phi(1)\, \phi(a)\, b' = \phi(a)\, b'$ $\quad [\therefore \phi(1)\, \phi(a) = \phi(a)]$

$\Rightarrow$ $\phi(a)\, [\phi(1)\, b'] = \phi(a)\, b'$

[$\because$ $\phi(1), \phi(a) \in R'$ which, being an integral domain, is a commutative ring]

$\Rightarrow \quad \phi(a)[\phi(1)b'] - \phi(a)b' = 0$

$\Rightarrow \quad \phi(a)[\phi(1)b' - b'] = 0$

$\Rightarrow \quad \phi(1)b' - b' = 0$

[$\because$ $\phi(a) \neq 0$ and R' is without zero divisors]

$\Rightarrow \quad \phi(1)b' = b' = b'\phi(1).$ [$\because$ R' is a commutative ring]

Thus $\quad \phi(1)b' = b'\phi(1) \ \forall \ b' \in R'.$

$\therefore$ $\phi(1)$ is the unit element of R'.

Example 3:

Show that every homomorphic image of a commutative ring is commutative.

Solution:

Let R be a commutative ring. Let f be a homomorphic mapping of R onto a ring R'. Then R' is a homomorphic image of R.

Let a', b' be any two elements of R'. Then f (a) = a', f (b) = b' for some a, b $\in$ R because f is onto R'. We have

$$a'b' = f(a)f(b) = f(ab)$$

$$= f(ba) \quad [\because \text{R is commutative}]$$

$$= f(b)f(a) = b'a'.$$

$\therefore$ R' is a commutative ring.

Example 4:

Prove that any homomorphism of a field is either an isomorphism or takes each element into 0.

Show that a field has no proper homomorphic image.

Solution:

Let ϕ be a homomorphism of a field F into a ring R. Let S be the kernel of ϕ. Then S is an ideal of the field F. We know that field has no proper ideal. Therefore either S = F or S = (0).

If S = F, then by definition of kernel of ϕ, we have $\phi(x) = 0 \ \forall \ x \in$ F. Thus, in this case ϕ takes each element of F into the zero element of R. In other words in this case ϕ (F) is the zero subring of the ring R.

If S = (0), then the kernel consists of zero element alone. So in this case ϕ is an isomorphism of F into R. Since the isomorphic image of a field is a field, therefore in this case ϕ (F) is a field isomorphic to the field F.

3.10 MAXIMAL IDEAL

An ideal S of ring (R, +.) is called maximal if there exist no ideal properly contained in R which properly contains in S that S is a maximal if for any other ideal T or R for which

$S \subseteq T \subseteq R$ *we have either*

$T = S$ *or* $T = R$.

An ideal $S \neq R$ *in a ring R is said to be a maximal ideal of R if whenever U is an ideal of R such that* $S \subseteq U \subseteq R$, *then either* $R = U$ *or* $S = U$.

In other words an ideal S of a ring R is said to be *maximal ideal* if there exists no ideal properly contained in R which itself properly contains S *i.e.*, if it is impossible to find an ideal which lies between S and the full ring R. For example, in the ring of integers I, the ideal (6) is not maximal since it is properly contained in I. On the other hand, (5) is a maximal ideal since the only ideal properly containing (5) is I itself.

Theorem:

A ideal S of the ring of integers **I** *is maximal if and only if S is generated by some prime integer.*

Proof:

We know that every ideal of the ring of integers **I** is a principal ideal. Suppose S is an ideal of **I** generated by P so that S = (p). Since p and – p generate the same ideal, therefore we can take p as positive.

Now are to prove that

(i) S is maximal if p is prime.

(ii) p is prime is S is maximal.

First we shall prove (i). Let p be a prime integer such that (p) = S. Let T be an ideal of I such that $S \subseteq T \subseteq \mathbf{I}$. Since T is also a principal ideal of **I**, let T = (q) where q is some positive integer.

Now $\quad S \subseteq T \Rightarrow (p) \subseteq T$

$\Rightarrow p \in T$

$\Rightarrow p \subseteq \{xq : x \in I\}$

$\Rightarrow p = rq$ for some positive integer r.

Since p is prime therefore q = 1 or q = p.

If q = 1, we have T = (q) = (1) = **I**

and if $\quad$ q = p, we have T = (q) = (p) = S.

Thus either T **I** or T = S.

Hence (p) is a maximal ideal of **I**.

Now we shall prove (ii). Let (p) = S be a maximal ideal. We are to show that p is prime. Let us suppose that p is a composite integer.

Let $p = mn$, $m \neq 1$, $n \neq 1$.

It is obvious that $(p) \subseteq (m) \subseteq$ **I**.

But since (p) is maximal ideal, therefore we have

either (m) = (p) or (m) = **I**.

If (m) = **I**, then m = 1, which is a contradiction.

If (m) = (p), then m must be equal to lp for some integer l since each element of (p) is a multiple of p.

Therefore $p = mn = lpn = pln$. But $p \neq 0$, therefore $ln = 1$. This gives $n = 1$ which is again a contradiction.

Hence p must be a prime integer.

3.11 SOME MORE RESULTS ON IDEALS

Theorem 1:

Let S_1, S_2 be ideals of a ring R and let

$$S_1 + S_2 = \{s_1 + s_2 : s_1 \in S_1, s_2 \in S_2\}.$$

Then $S_1 + S_2$ is an ideal of R generated by $S_1 \cup S_2$.

Proof:

Let $a_1 + a_2 \in S_1 + S_2$, $b_1 + b_2 \in S_1 + S_2$. Then

$a_1, b_1 \in S_1$ and $a_2, b_2 \in S_2$.

We have $(a_1 + a_2) - (b_1 + b_2) = (a_1 - b_1) + (a_2 - b_2)$.

Since S_1 is an ideal, therefore $a_1, b_1 \subset S_1 \Rightarrow a_1 - b_1 \in S_1$.

Similarly $a_2 - b_2 \subset S_2$.

$\therefore (a_1 - b_1) + (a_2 - b_2) \in S_1 + S_2$.

$\therefore (a_1 + a_2) - (b_1 + b_2) \in S_1 + S_2$.

$\therefore S_1 + S_2$ is a subgroup of the additive group of R

Let r be any element of R. Then

$r(a_1 + a_2) = ra_1 + ra_2 \in S_1 + S_2$ since $r \in R$, $a_1 \in S_1$

$\Rightarrow ra_1 \in S_1$ and similarly $ra_2 \in S_2$.

Similarly $(a_1 + a_2)\, r = a_1 r + a_2 r \in S_1 + S_2$ since $a_1 r \in S_1$, $a_2 r \in S_2$.

Hence $S_1 + S_2$ is an ideal of R.

Since $0 \in S_1$ and also $0 \in S_2$, therefore obviously

$S_1 \subseteq S_1 + S_2$ and $S_2 \subseteq S_1 + S_2$.

$\therefore S_1 \cup S_2 \subseteq S_1 + S_2$.

Thus $S_1 + S_2$ is an ideal of R containing $S_1 \cup S_2$.

Also if S is any ideal of R containing $S_1 \cup S_2$, then S must contain $S_1 + S_2$. Thus $S_1 + S_2$ is the smallest ideal of R containing $S_1 \cup S_2$ *i.e.*, $S_1 + S_2 = (S_1 \cup S_2)$.

Theorem 2:

If an ideal U of a ring R contains a unit of R then U = R.

Proof:

Let R be a ring with unity element 1. Let u be a unit of R. Then u is an inversible element of R *i.e.*, u^{-1} exists. Let $u \in U$.

Since U is an ideal, therefore

$u \in U, u^{-1} \Rightarrow uu^{-1} \in U \Rightarrow 1 \in U$.

Now let x be any element of R. Then

$x \in R, 1 \in U \Rightarrow x\,1 \in U \Rightarrow x \in U$.

$\therefore R \subseteq U$.

Also $U \subseteq R$ as U is an ideal of R. Hence $R \neq U$.

Theorem 3:

Let R be a commutative ring with unity and a, b be two non-zero elements of R. Then

(a) = (b) iff a | b and b | a.

Proof:

Here (a) = the principal ideal of R generated by a

$= \{ax : x \in R\}$.

Similarly (b) = the principal ideal of R generated by b.

Let (a) = (b).

Then $(a) \subseteq (b)$

$\Rightarrow a \in (b)$

$\Rightarrow a = rb$ for some $r \in R$

$\Rightarrow$ b | a *i.e.*, b is a divisor of a.

Similarly (a) = (b)

$\Rightarrow$ (b) $\subseteq$ (a) $\Rightarrow$ b $\in$ (a)

$\Rightarrow$ b = sa for some s $\in$ R $\Rightarrow$ a | b.

Thus (a) = (b) $\Rightarrow$ a | b and b | a.

Conversely let a | b and b | a.

Now a | b $\Rightarrow$ b = pa for some p $\in$ R. Let y be any element of (b).

Then y = ub for some u $\in$ R

= u (pa) = (up) a $\in$ (a) since up $\in$ R.

Thus y $\in$ (b) $\Rightarrow$ y $\in$ (a)

$\therefore$ (b) $\subseteq$ (a).

Thus a | b $\Rightarrow$ (b) $\subseteq$ (a). Similarly b | a $\Rightarrow$ (a) $\subseteq$ (b).

Consequently a | b, b | a $\Rightarrow$ (a) = (b).

Corollary: *Let R be an integral domain with unity and a, b be two non-zero elements of R. Then (a) = (b) iff a and b are associates.*

Theorem 4:

An ideal S of a commutative ring R with unity is maximal if and only if the residue class ring R/S is a field.

Proof:

Since R is a commutative ring with unity, therefore R/S is also a commutative ring with unity. The zero element of the ring R/S is S and the unity element is the coset S + 1 where 1 is the unity element of R.

Let the ideal S be maximal. Then to prove that R/S is field.

Let S + b be any non-zero element of R/S. Then S + b $\neq$ S *i.e.*, b $\notin$ S. To prove that S + b is inversible.

If (b) is the principal ideal of R generated by b, then S + (b) is also an ideal of R. Since b $\notin$ S, therefore the ideal S is properly contained in S + (b). But S is a maximal ideal of R. Hence we must have S + (b) = R.

Since 1 $\in$ R, therefore we must obtain 1 on adding an element of S to an element of (b). Therefore there exists an element a $\in$ S and $\alpha \in$ R such that

a + αb = 1 [Note that (b) = {αb: $\alpha \in$ R}]

$\therefore$ 1 – ab = a $\in$ S.

Consequently $S + 1 = S + \alpha b = (S + \alpha)(S + b)$

$\therefore S + \alpha = (S + b)^{-1}$. Thus S + b is inversible.

$\therefore$ R/S is a field.

Conversely, let S be an ideal of R such that R/S is a field. We shall prove that S is a maximal ideal of R.

Let S' be an ideal of R properly containing S *i.e.*, $S \subseteq S'$ and $S \neq S'$. Then S will be maximal if S' = R. The elements of R contained in S already belong to S' since $S \subseteq S'$. Therefore R will be subset of S' if every element α of R not contained in S also belongs to S'. If $\alpha \in$ is such that $\alpha \notin S$, then $S + \alpha \neq S$ *i.e.*, $S + \alpha$ is a non-zero element of R/S. Also S' properly contains S. Therefore there exists an element β of S' not contained in S so that $S + \beta$ is also a non-zero element of R/S. Now the non-zero elements of R/S form a group with respect to multiplication because R/S is a field. Therefore there exits a non-zero element S + y of R/S such that

$$(S + y)(S + \beta) = S + \alpha.$$

[We may take $S + y = (S + \alpha)(S + \beta)^{-1}$].

Now $(S + y)(S + \beta) = S + \alpha$

$\Rightarrow S + y\beta = S + \alpha \Rightarrow y\beta - \alpha \in S \Rightarrow y\beta - \alpha \in S'$. $\quad [\because S \subseteq S']$

Now S' is an ideal. Therefore $y \in R, \beta \in S' \Rightarrow y\beta \in S'$.

Again $y\beta \in S', y\beta - \alpha \in S' \Rightarrow y\beta - (y\beta - \alpha) \in S'$ *i.e.*, $\alpha \in S'$.

Thus $R \subseteq S'$. Also $S' \subseteq R$ as S' is an ideal of R.

$\therefore S' = R$.

Hence the theorem.

2.12 PRIME IDEALS

Prime Ideal

Definition: *Let R be a ring and S an ideal in R. Then S is said to be a prime ideal of R if ab $\in$ S, a, b $\in$ R implies that either a or b is in S.*

For example in the ring of integers I, the principal ideal (7) is prime. Obviously if ab is in (7), then a or b must be a multiple of 7. On the other hand, (6) is not a prime ideal in since, in particular, $12 = 3 \times 4$ is in (6), yet neither 3 nor 4 is an element of (6).

Theorem 1:

Let R be a commutative ring and S an ideal of R. Then the ring of residue classes R/S is an integral domain if and only if S is a prime ideal.

Proof:

Let R be a commutative ring and S an ideal of R. Then

$$R/S = \{S + a: a \in R\}.$$

Let S + a, S + b be any two elements of R/S. Then $a, b \in R$.

We have $(S + a)(S + b) = S + ab$

$= S + ba$ [$\because$ R is a commutative ring]

$= (S + b)(S + a)$.

$\therefore$ R/S is a commutative ring.

Now let S be a prime ideal of R. Then we are to prove that R/S is an integral domain. For this we are to show that R/S is without zero divisors. The zero element of the ring R/S is the residue class S itself. Let S + a, S + b be any two elements of R/S.

Then $(S + a)(S + b) = S$ (the zero element of R/S)

$\Rightarrow S + ab = S \Rightarrow ab \in S$

$\Rightarrow$ either a or b is in S, since S is a prime ideal

$\Rightarrow$ either $S + a = S$ or $S + b = S$ [Note that $a \in S \Leftrightarrow S + a = S$]

$\Rightarrow$ either S + a or S + b is the zero element of R/S.

$\therefore$ R/S is without zero divisors.

Since R/S is a commutative ring without zero divisors, therefore

R/S is an integral domain.

Conversely, let R/S be an integral domain. Then we are to prove that S is a prime ideal of R. Let a, b be any two elements in R such that $ab \in S$. We have

$$ab \in S \Rightarrow S + ab = S \Rightarrow (S + a)(S + b) = S.$$

Since R/S is an integral domain, therefore it is without zero divisors. Therefore

$(S + a)(S + b) = S$ (the zero element of R/S)

$\Rightarrow$either S + a or S + b is zero $\Rightarrow$ either $S + a = S$ or $S + b = S$

$\Rightarrow$either $a \in S$ or $b \in S \Rightarrow$ S is a prime ideal.

This completes the proof of the theorem.

Theorem 2:

Let R be a commutative ring with unity. Then every maximal ideal of R is a prime ideal.

Proof:

R is a commutative ring with unit element. Let S be a maximal ideal of R. Then R/S is a field.

Now every field is an integral domain. Therefore R/S is also an integral domain. Hence by theorem 1, S is a prime ideal of R. This completes the proof of the theorem.

But it should be noted that the converse of the above theorem is not true *i.e.*, every prime ideal is not necessarily a maximal ideal.

Example 1:

If R is a finite commutative ring (i.e. has only a finite number of elements) with unit element prove that every prime ideal of R is a maximal ideal of R.

Solution:

Let R be a finite commutative ring with unit element. Let S be a prime ideal of R. Then to prove that S is a maximal ideal of R.

Since S is a prime ideal of R, therefore the residue class ring R/S is an integral domain. Now

$$R/S = \{S + a : a \in R\}.$$

Since R is a finite ring, therefore R/S is a finite integral domain. But every finite integral domain is a field. Therefore R/S is a field. Since R is a commutative ring with unity and R/S is a field, therefore S is a maximal ideal of R.

Example 2:

Let R be the ring of all real valued continuous functions defined on the closed interval [0, 1]. Let

$$M = \{f(x) \in R : f\left(\frac{1}{3}\right) = 0\}.$$

Show that M is a maximal ideal of R.

Solution:

First of all we observe that M is non empty because the real valued function e (x) on [0, 1] defined by

$$e(x) = 0 \ \forall \ x \in [0, 1]$$

belongs to M.

Now let f (x), g (x) be any two elements of M. Then

$f\left(\frac{1}{3}\right) = 0$, $g\left(\frac{1}{3}\right) = 0$, by definition of M.

Let h (x) = f (x) – g (x).

Then $h\left(\frac{1}{3}\right) = f\left(\frac{1}{3}\right) - g\left(\frac{1}{3}\right) = 0 - 0 = 0$.

Therefore h (x) ∈ M.

Thus f (x), g (x) ∈ M ⇒ h (x) = f (x) – g (x) ∈ M.

Further let f (x) be any element of M and r (x) be any element of R.

Then $f\left(\frac{1}{3}\right) = 0$, by definition of M.

Let t (x) = r (x) f (x) = f (x) r (x). [∴ R is a commutative ring].

Then $t\left(\frac{1}{3}\right) = r\left(\frac{1}{3}\right) f\left(\frac{1}{3}\right) = r\left(\frac{1}{3}\right) . 0 = 0$. Therefore t (x) ∈ M.

Thus r (x) ∈ R, f (x) ∈ M ⇒ r (x) f (x) ∈ M.

Hence M is an ideal of R.

Clearly M ≠ R because i (x) ∈ R given by i (x) = 1 ∀ x ∈ [0, 1] does not belong to M.

The ring R is with unity and the element i (x) is its unity element.

Let N be an ideal of R properly containing M *i.e.*, M ∈ N and M ≠ N. Then M will be a maximal ideal of R if N = R, which be so if the unity i (x) of R belongs to N. Since M is a proper subset of N, therefore there exists λ (x) ∈ N such that λ (x) ∉ M. This means $\lambda\left(\frac{1}{3}\right) \neq 0$. Put $\lambda\left(\frac{1}{3}\right)$ = c where c ≠ 0.

Let us define β (x) ∈ R, by β (x) = c ∀ x ⊂ [0, 1]. Now consider μ (x) ∈ R given by μ (x) = λ (x) – β (x).

We have $\mu\left(\frac{1}{3}\right) = \lambda\left(\frac{1}{3}\right) - \beta\left(\frac{1}{3}\right) = c - c = 0$.

Therefore μ (x) ∈ M and so μ(x) also belongs to N because N is a superset of M. Now N is an ideal of R and λ(x), μ (x) are in N. Therefore λ(x) – μ(x) = β(x) is also an element of N.

Now define $\lambda(x) \in R$ by $\lambda(x) = 1/c \ \ \forall \ x \in [0, 1]$. Since N is an ideal of R, therefore $\lambda(x) \in R$ and $\beta(x) \in N \Rightarrow \lambda(x) \beta(x) \in N$. We shall show that $\lambda(x) \beta(x) = i(x)$.

For every $x \in [0, 1]$, we have

$$\lambda(x) \beta(x) = (1/c)c = 1.$$

Therefore $\lambda(x) \beta(x) = i(x)$, by definition of i(x).

Thus the unity element i(x) of R belongs to N and consequently $N = R$.

Hence M is a maximal ideal of R.

Example 3:

Give an example of a ring in which some prime ideal is not a maximal ideal.

Solution:

Let us consider **I** [x] be the ring of polynomials over the ring of integers **I**. Let S be the principal ideal of **I** [x] generated by x *i.e.*, let S = (x). We shall show that (x) is prime but not maximal.

We have $S = (x) = \{x f(x) : f(x) \in \mathbf{I}[x]\}$.

Now we shall prove that S is prime.

Let $a(x), b(x) \in \mathbf{I}[x]$ be such that $a(x) b(x) \in S$. Then there exists a polynomial $c(x) \in \mathbf{I}[x]$ such that

$$x\, c(x) = a(x)\, b(x). \qquad \text{...(1)}$$

Let $a(x) = a_0 + a_1x + a_1x + a_2x^2 + \ldots$, $b(x) = b_0 + b_1x + b_2x^2 + \ldots$,

$c(x) = c_0 + c_1x + c_2x^2 + \ldots$. Then (1) becomes

$$x(c_0 + c_1x + \ldots) = (a_0 + a_1x + \ldots)(b_0 + b_1x + \ldots).$$

Equation the constant term on both sides, we obtain

$$a_0b_0 = 0$$

$$\Rightarrow a_0 = 0 \text{ or } b_0 = 0. \qquad [\because \text{ I is without zero divisors}]$$

Now $a_0 = 0 \Rightarrow a(x) = a_1x + a_2x^2 + \ldots$

$\Rightarrow a(x) = x(a_1 + a_2x + \ldots) \Rightarrow a(x) \in (x)$.

Similarly $b_0 = 0 \Rightarrow b(x) = b_1x + b_2x^2 + \ldots$

$\Rightarrow b(x) = x(b_1 + b_2x + \ldots) \Rightarrow b(x) \in (x)$.

Thus $a(x) b(x) \in (x) \Rightarrow$ either $a(x) \in (x)$ or $b(x) \in (x)$.

Hence (x) is a prime ideal.

Now to prove that (x) is not a maximal ideal of **I** [x]. For this we must show an ideal N of **I** [x] such that (x) is properly contained in N, while N itself is properly contained in **I** [x]. The ideal N = (x, 2) serves this purpose.

Obviously $(x) \subseteq (x, 2)$. In order to show that (x) is properly contained in (x, 2) we must show an element of (x, 2) which is not in (x). Clearly $2 \in (x, 2)$. We shall show that $2 \notin (x)$. Let $2 \in (x)$. Then we can write,

$$2 = x\, f(x) \text{ for some } f(x) \in \mathbf{I}[x].$$

Let $f(x) = a_0 + a_1 x + \ldots$

Then $2 = x\, f(x) \Rightarrow = x\,(a_0 + a_1 x + \ldots)$

$\Rightarrow 2 = a_0 x + a_1 x^2 + \ldots$

$\Rightarrow 2 = 0 + a_0 x + a_1 x^2 + \ldots$

$\Rightarrow 2 = 0$ [by equality of two polynomials]

But $2 \neq 0$ in the ring of integers. Hence $2 \notin (x)$. Thus (x) is properly contained in (x, 2).

Now obviously $(x, 2) \subseteq \mathbf{I}[x]$. In order to show that (x, 2) is properly contained in **I** [x] we must show an element of **I** [x] which is not in (x, 2). Clearly $1 \in \mathbf{I}[x]$. We shall show that $1 \notin (x, 2)$. Let $1 \in (x, 2)$. Then we have a relation of the form

$$1 = x\, f(x) + 2\, g(x), \quad \text{where } f(x), g(x) \in \mathbf{I}[x].$$

Let $f(x) = a_0 + a_1 x + \ldots\; g(x) = b_0 + b_1 x + \ldots$

Then $1 = x\,(a_0 + a_1 x + \ldots) + 2\,(b_0 + b_1 x + \ldots)$

$\Rightarrow 1 = 2\, b_0$ [Equating constant term on both sides]

But there is no integer b_0 such that $1 = 2\, b_0$.

Hence $1 \notin (x, 2)$. Thus (x, 2) is properly contained in **I** [x].

Therefore (x) is not a maximal ideal of **I** [x]

Example 4:

Let **R** *be the field of real numbers and S the set of all those polynomials* $f(x) \in \mathbf{R}[x]$ *such that* $f(0) = 0 = f(1)$. *Prove that S is an ideal of* **R** [x]. *Is the residue class ring* **R** [x]/S *an integral domain? Given reasons for your answer.*

Solution:

Let f (x), g (x) be any elements of S. Then

$f(0) = 0 = f(1)$ any $g(0) = 0 = g(1)$.

Let $h(x) = f(x) - g(x)$. Then

$h(0) = f(0) - g(0) = 0 - 0 = 0$ and $h(1) = f(1) - g(1) = 0 - 0 = 0$.

Thus $h(0) = 0 = h(1)$. Therefore $h(x) \in S$.

Thus $f(x), g(x) \in S \Rightarrow h(x) = f(x) - g(x) \in S$.

Further let $f(x)$ be any element of S and $r(x)$ be any element of **R** [x]. Then $f(0) = 0 = f(1)$, by definition of S.

Let $t(x) = r(x) f(x) = f(x) r(x)$. [$\because$ **R** [x] is a commutative ring]

Then $t(0) = r(0) f(0) = r(0).0 = 0$

and $t(1) = r(1) f(1) = r(1).0 = 0$.

$\therefore$ $t(x) \in S$.

Thus $r(x) \in$ **R** [x], $f(x) \in S \Rightarrow r(x) f(x) \in S$.

Hence S is an ideal of **R** [x].

Now we claim that S is not a prime ideal of **R** [x]. Let $f(x) = x(x - 1)$. Then $f(0) = 0(0 - 1) = 0$, and $f(1) = 1(1 - 1) = 0$.

Thus $f(x) = x(x - 1)$ is an element of S.

Now let $p(x) = x$, $q(x) = x - 1$.

We have $p(1) = 1 \neq 0$. Therefore $p(x) \notin S$.

Also $q(0) = 0 - 1 = -1 \neq 0$.

Therefore $q(x) \notin S$. Thus $x(x - 1) \in S$ while neither $x \in S$ nor $x - 1 \in S$. Hence S is not a prime ideal of **R** [x].

Since S is not a prime ideal of **R** [x], therefore the residue class ring **R** [x]/S is not an integral domain.

3.13 EUCLIDEAN RINGS OR EUCLIDEAN DOMAINS

Definition: *Let R be an integral domain i.e., let R be a commutative ring without zero divisors. Then R is said to be a Euclidean ring if to every non-zero element $a \in R$ we can assign a non-negative integer d (a) such that :*

(i) For all $a, b \in R$, both non-zero, $d(ab) \geq d(a)$.

(ii) For any $a, b \in R$ and $b \neq 0$, there exist $q, r \in R$ such that $a = qb + r$ where either $r = 0$ or $d(r) < d(b)$.

The second part of the above definition is known as division *algorithm.* Also we do not assign a value of d (0). Thus d (a) will remain undefined when $a = 0$. Also d (a) will be called d-value of a and d (a) must be sone non-negative integer for every non-zero element $a \in R$.

Example 1:

The ring of Gaussian integers is a Euclidean ring.

Solution:

Let (G, +,.) be the ring of Gaussian integers where

$$G = \{x + iy: x,y \in \mathbf{I}\}.$$

Let the d function on the non-zero elements of G be defined as

$$d\,(x + iy) = x^2 + y^2 \ \forall\ 0 + i0 \neq x + iy \in G.$$

Now if x + iy is a non-zero element of G, then $(x^2 + y^2)$ is a non-negative integer. Thus we have assigned a non-negative integer to every non-zero element of G.

If x + iy and m + in are two non-zero elements of G, then

$$d\,[(x + iy)\,(m + in)] = d\,[(xm - ny) + i\,(my + xn)]$$

$$= (xm - ny)^2 + (my + xn)^2 = x^2m^2 + n^2y^2 + m^2y^2 + x^2n^2$$

$$= (x^2 + y^2)\,(m^2 + n^2)$$

$$\geq x^2 + y^2. \qquad [\because\ m^2 + n^2 \geq 1]$$

Thus $d\,[x + iy)\,(m + in)] \geq d\,(x + iy)$.

Now to show the existence of division algorithm in G.

Let $\alpha \in G$ and let β be a non-zero element of G. Let $\alpha = x + iy$ and $\beta + m + in$. Define a complex number λ be the equation

$$\lambda = \frac{\alpha}{\beta} = \frac{x + iy}{m + in} = \frac{(x + iy)\,(m - in)}{m^2 + n^2} = p + iq$$

where p, q are rational numbers.

Here λ is not necessarily a Gaussian integer.

Also division by β is possible since $\beta \neq 0$.

Let p' and q' be the nearest integers to p and q respectively.

Then obviously $|p - p'| \leq \frac{1}{2}$, $|q - q'| \leq \frac{1}{2}$.

Let $\lambda' = p' + iq'$. Then λ' is a Gaussian integer.

Now $\lambda = \frac{\alpha}{\beta} \Rightarrow \alpha = \lambda\beta$

$$\Rightarrow \alpha = \lambda'\beta + \lambda\beta - \lambda'\beta.$$

Thus $\alpha = \lambda'\beta + (\lambda - \lambda')b$. ...(1)

Since α, β, γ' are Gaussian integers, therefore from (1) it implies that $(\lambda - \lambda')\,\beta$ is also a Gaussian integer.

Now if p and q are integers then $p = p'$, $q = q'$.

So $\lambda - \lambda' = (p - p') + i\,(q - q') = 0 + i0$. Thus $(\lambda - \lambda')\,\beta = 0 + i0$.

If p and q are not both integers, then $(\lambda - \lambda')\,\beta$ is a non-zero Gaussian integer and we have

$$d\,[(\lambda - \lambda')\beta] = d\,[\{(p - p') + i\,(q - q')\}\,(m + in)]$$

$$= [(p - p')^2 + (q - q')^2]\,(m^2 + n^2) = [(p - p')^2 + (q - q')^2]\,d\,(\beta)$$

$$\leq \left[\frac{1}{4} + \frac{1}{4}\right] d\,(\beta) \qquad \left[\because\ (p - p')^2 \leq \frac{1}{4},\ (q - q')^2 \leq \frac{1}{4}\right]$$

$$= \frac{1}{2}\,d\,(\beta) < d\,(\beta).$$

Thus $\alpha = \lambda'\beta\,(\lambda - \lambda')\,\beta$ where λ' and $(\lambda - \lambda')\,\beta$ are Gaussian integers either $(\lambda - \lambda')\,b = 0$

$$\Rightarrow \qquad d\,[(\lambda - \lambda')\,b] < d\,(\beta).$$

Hence the ring of Gaussian integers is a Euclidean ring.

Example 2:

The ring of polynomials over field is a Euclidean ring.

Solution:

Let F [x] be the ring of polynomials over a field F. Let the d function on the non-zero polynomials in F [x] be defined as

$$d\,[f\,(x) = \deg f\,(x),\ \forall\ 0 \neq f\,(x) \in F\,[x].$$

Now if $0 \neq f\,(x) \in F\,[x]$, then deg f (x) is a non-negative integer.

Thus we have assigned a non-negative integer to every non-zero element f (x) in F [x].

Further if f (x), g (x) $\in$ F [x] and are both non-zero polynomials, then

$$\deg\,[f\,(x)] = \deg f\,(x) + \deg g\,(x)$$

$$\Rightarrow \qquad \deg\,[f\,(x)\,g\,(x) \geq \deg f\,(x) \qquad [\because\ \deg g\,(x) \geq 0]$$

$$\Rightarrow \qquad d\,[f\,(x)\,g\,(x) \geq d\,[f\,(x)].$$

Finally we know that if $f\,(x) \in F\,[x]$ and $0 \neq g\,(x) \in F\,[x]$, then there exist two polynomials q (x) and r (x) in F [x] such that

$$f\,(x) = q\,(x)\,g\,(x) + r\,(x)$$

where either r (x) = 0 or deg r (x) < deg g (x)

i.e. where either r (x) = 0 or d [r (x)] < d [g (x)].

Hence the ring of polynomials over a field is a Euclidean ring.

Example 3:

The ring of integers is a Euclidean ring.

Solution:

Let (**I**, +,.) be the ring of integers where

$\mathbf{I} = \{..., -3, -2, -1, 0, 1, 2, 3,....\}$.

Let the d function on the non-zero elements of **I** be defined as

$d(a) = |a| \ \forall\ 0 \neq a \in \mathbf{I}$.

Now if $0 \neq a \in \mathbf{I}$, then $|a|$ is a non-negative integer. Thus we have assigned a non-negative integer to every non-negative element $a \in \mathbf{I}$.

[$d(-5) = |-5| = 5$, $d(-1) = |-1| = 1$, $d(4) = |4| = 4$ etc.]

Further if a, b ∈ **I** and are both non-zero, then

$|ab| = |a||b|$

$\Rightarrow |ab| \geq |a|$ [$\because |b| \geq 1$ if $0 \neq b \in \mathbf{I}$]

Finally we known that if $a \in \mathbf{I}$ and $0 \neq b \in \mathbf{I}$, then there exist two integers q and r such that

$$a = qb + r \text{ where } 0 \leq r < |b|$$

i.e., where either r = 0 or $\leq r < |b|$

i.e., where either r = 0 or d (r) < d (b).

It should be noted that d (b) = | b | and if r is a positive integer then $r = |r| = d(r)$.

Hence the ring of integers is a euclidean ring.

Example 4:

Every field is a Euclidean ring.

Solution:

Let F be any field. Let the d function on the non-zero elements of F be defined as

$$d(a) = \forall\ 0 \neq a \in F.$$

Thus we have assigned the integer zero to every non-zero element in F.

If a and b are non-zero elements in F then ab is also a non-zero element in F. We have therefore

$$d(ab) = 0 = d(a).$$

Thus we have $d(ab) \geq d(a)$.

Finally if $a \in F$ and $0 \neq b \in F$, then we can write

$$a = (ab^{-1})\, b + 0$$

i.e., $a = qb + r$ where $q = q = ab^{-1}$ and $r = 0$.

Hence every field is a Euclidean ring.

Theorem

Every Euclidean ring possesses unity element.

Proof:

Let R be a Euclidean ring. Obviously R is an ideal of R. Therefore there exists an element $u_0 \in R$ such that $R = (u_0)$ *i.e.*, there exists an element $u_0 \in R$ such that every element in R is a multiple of u_0. Since, in particular, $u_0 \in R$ therefore there exists an element $c \in R$ such that $u_0 = u_0c$. We shall show that c is the required unity element. Let now a be any element of R. Since $a \in R$, therefore there exists some $x \in R$ such that $a = u_0x$.

Now $ac = (u_0x)\, c$ $[\because a = u_0x]$

$= (u_0c)\, x$ $[\because$ R is a commutative ring$]$

$= u_0x$ $[\because u_0 = u_0c]$

$= a.$ $[\because a = u_0x]$

Thus we have $ac = a = ca \ \forall \ a \in r$.

Hence c is the unity element.

3.14 PROPERTIES OF EUCLIDEAN RINGS

Theorem 1:

Every Euclidean ring is a principal ideal ring.

Proof:

Let R be a Euclidean ring. Let S be an arbitrary ideal of R. If S is the null ideal, then $S = (0)$ *i.e.*, the ideal of R generated by 0. Therefore S is a principal ideal. So let us suppose that S is not a null ideal. Then there exist elements in S not equal to zero. Let b be a non-zero element in S such that $d(b)$ is minimum *i.e.*, there exists no element c in S such that $d(c) < d(b)$. We shall show that $S = (b)$ *i.e.*, S is nothing but the ideal generated by b.

Let a be any element of S. Then by definition of Euclidean ring there exist elements q and r in R such that

$$a = qb + r \text{ where either } r = 0 \text{ or } d(r) < d(b).$$

Now $q \in R$, $b \in S \Rightarrow qb \in S$ because S is an ideal.

Further $a \in S$, $qb \in S \Rightarrow a - qb = r \in S$.

Thus $r \in S$ and we have either $r = 0$ or $d(r) < d(b)$.

If $r \neq 0$, then $d(r) < d(b)$ which contradicts our assumption that no element in S has d-value smaller than d (b). Therefore we must have $r = 0$.

Then $\qquad a = qb.$

Thus every element a in S is a multiple of the generating element b. Thus $a \in S \Rightarrow a \in (b)$. Therefore $S \subseteq (b)$.

Again if xb is any element of (b), then $x \in R$.

Now $x \in R$, $b \in S \Rightarrow xb \in S$. Therefore $(b) \subseteq S$.

Hence $\qquad S = (b).$

Thus, every ideal S in R is a principal ideal. Therefore R is a principal ideal ring.

Theorem 2:

The necessary and sufficient condition that the non-zero element in the Euclidean ring R is a unit is that

$$d(a) = d(1).$$

Proof:

Let a be a unit in R. Then to prove that $d(a) = d(1)$.

By the definition of Euclidean ring

$$d(1a) \geq d(1)$$

$$\Rightarrow \qquad d(a) \geq d(1). \qquad \text{...(1)}$$

Since a is a unit in R, therefore a^{-1} exists and we have

$$1 = aa^{-1}$$

$$\Rightarrow \qquad d(1) = d(aa^{-1}).$$

But $\qquad d(aa^{-1}) \geq d(a).$

$$\therefore \qquad d(1) \geq d(a). \qquad \text{...(2)}$$

Conversely, let $d(a) = d(1)$. Then to prove that a is a unit in R. If a is not a unit in R, then by theorem 6, we have

$$d(1a) > d(1).$$

$\Rightarrow \quad d(a) > d(1).$

Thus we get a contradiction. Hence a must be a unit in R.

Theorem 3:

An ideal S of the Euclidean ring R is maximal iff S is generated by some prime element of R.

Proof:

We know that every ideal of a Euclidean ring R is a principal ideal. Suppose S is an ideal of R generated by p so that S = (p). Now we are to prove that

(i) S is maximal if p is a prime element of R.

(ii) p is prime if S is maximal.

First we shall prove (i). Let p be a prime element of R such that (p) = S. Let T be an ideal of R such that $S \subseteq T \subseteq R$. Since T is also a principal ideal of R, so let T = (q) where $q \in R$.

Now $\quad S \subseteq T \Rightarrow (p) \subseteq (q) \Rightarrow p \in (q)$

$\Rightarrow \quad p = xq$ for some $x \in R \Rightarrow q \mid p$.

Since p is prime, therefore either q should be a unit in R or q should be an associate of P.

If q is a unit in R, then T = (q) = R.

If q is an associate of p, then T = (q) = (p) = S.

Thus either T = R or T = S.

Now we shall prove (ii). Let (p) = S be a maximal ideal. We are to show that p is prime. Let us suppose that p is composite *i.e.*, p is not prime.

Let p = mn where neither m nor n is unit in R.

Now $p = mn \Rightarrow m \mid p$

$\Rightarrow (p) \subseteq (m).$

But $(m) \subseteq R$. Therefore we have $(p) \subseteq (m) \subseteq R$.

But (p) is a maximal ideal, therefore we should have either

(m) = (p) or (m) = R.

If R = (m), then $R \subseteq (m)$.

$\therefore \quad 1 \in R \Rightarrow 1 \in (m)$

$\Rightarrow$ $1 = ym$ for some $y \in R$

$\Rightarrow$ m is inversible $\Rightarrow$ m is a unit in R.

Thus we get a contradiction.

If $(m) = (p)$, then $m \in (p)$. Therefore $m = lp$ for some $l \in R$.

$\therefore p = mn = lpn = pln$.

$\therefore p(1 - ln) = 0$

$\Rightarrow 1 - ln = 0$ [$\because p \neq 0$ and R is without zero divisors]

$\Rightarrow ln = 1 \Rightarrow$ n inversible $\Rightarrow$ n is a unit in R.

Theorem 4:

Let R be a Euclidean ring and a and be any two elements R, not both of which are zero. Then a and b have a greatest common divisor d which can be expressed in the form

$d = \lambda a + \mu b$ for some $\lambda, \mu \in R$.

Proof:

Consider the set

$S = \{sa + tb: s, t \in R)$. ...(1)

We claim that S in an ideal of R. The proof is as follows:

Let $x = s_1a + t_1b$, and $y = s_1a + t_2b$ be any two elements of S.

Then $s_1, t_1, s_2, t_2 \in R$. We have

$x - y = (s_1a + t_1b) - (s_2a + t_2b) = (s_1 - s_2) a + (t_1 - t_2) b \in S$

since $s_1 - s_2$ and $t_1 - t_2$ are both elements of R.

Thus S is a subgroup of R with respect to addition.

Also if u be any element of R, then

$xu + ux = u(s_1a + t_1b) = (us_1) a + (ut_1) b \in S$ since $us_1, ut_1 \in R$.

Therefore S is an ideal of R. Row every ideal in R is a principal ideal. Therefore there exists an element d in S such that every element in S is a multiple of d.

Since $d \in S$, therefore from (1), we see that there exist elements $\lambda, \mu \in R$ such that $d = \lambda a + \mu b$

Now R is a ring with unity element 1.

$\therefore$ Putting $s = 1, t = 0$ in (1), we see that $a \in S$. Also putting $s = 0$, $t = 1$ in (1), we see that $b \in S$.

Now a, b are elements of S. Therefore they are both multiples of d. Hence d | a and d | b.

Now suppose c | a and c | b.

Then c | λa and c | μb. Therefore c is also a divisor of λa + μb *i.e.*, c is a divisor of d.

Thus d is a greatest common divisor of a and b.

Unique Factorization

Theorem 5:

Let R be a Euclidean ring and a be a non-zero non-unit element in R. Suppose that

$$a = p_1 p_2 \cdots p_m = q_1\, q_2 \cdots q_n.$$

where the p's and q's are prime elements of R. Then m = n and each p is an associate of some q and each q is an associate of some p.

Proof:

We have $p_1p_2 \ldots p_m = q_1q_2 \ldots q_n$. Now p_1 is a divisor of $p_1p_2 \ldots p_m$. Therefore p_1 is also a divisor of $q_1q_2 \ldots q_n$. By Cor. to Theorem. 5, p_1 must divide at least one of $q_1, q_2, \ldots, q_n$. Since R is a commutative ring, therefore without loss of generality we may suppose that p_1 divides q_1. But p_1 and q_1 are both prime elements in R. Therefore p_1 and q_1 must be associates and we have $q_1 = up_1$ where u is a unit in R. Thus we have

$$p_1p_2 \cdots p_m = up_1\, q_2 \cdots q_n.$$

Cancelling $0 \neq p_1$ from both sides, we get

$$p_2p_3 \cdots p_m = uq_2\, q_3 \cdots q_n. \qquad \ldots(1)$$

Now we can repeat the above argument on the relation (1) with ring p_2. If n > m, then after m steps the left hand side becomes 1 while the right hand side reduces to a product of some units in R and certain number of q's (the excess of n over m). But the q's are prime elements of R and so they are not units in R. So the product of some units in R and certain number of q's cannot be equal to 1. Therefore n cannot be greater than m.

Thus $n \leq m$.

Similarly interchanging the roles of p's and q's we get

$m \leq n$. Hence $m = n$.

Also in the above process we have shown that every p is an associate of some q and conversely ever q is an associate of some p. Hence the theorem has been completely established.

Note: Combining theorems 8 and 9, we can say that *every non-zero element in a Euclidean ring R can be uniquely written (upto associates) as a product of prime elements or is a unit in R. Therefore a Euclidean ring is a Unique Factorization Domain.*

Theorem 6:

If p is a prime element in the Euclidean ring R and p | ab where a, b ∈ R then p divides at least one of a or b.

Proof:

If p divides a, we are nothing to prove. So suppose that p does not divide a. Since p is prime and p does not divide a, therefore p and a are relatively prime *i.e.*, the greatest common divisor of p and a is 1. Hence by theorem 4, we get that p | b.

Corollary: *If p is a prime element in the Euclidean ring R and p divides the product $a_1\ a_2 \ldots a_n$ of elements in R, then p divides at least one of a_1, $a_2, \ldots, a_n$.*

The result follows immediately by repeated application of theorem.

Theorem 7:

Let R be a Euclidean ring. Let a and b be two non-zero elements in R. Then

(i) if b is a unit in R, d (ab) = d (a).

(ii) if b is not a unit in R, d (ab > d (a).

Proof:

(i) By the definition of Euclidean ring, we have

$$d\ (ab) \geq d\ (a). \qquad \ldots(1)$$

Now suppose that b is a unit in R. Then b is inversible and b^{-1} exists. We can write

$$a = (ab)\ b^{-1}.$$

$$\therefore \quad d\ (a) = d\ [(ab)\ b^{-1}]$$

But by the definition of Euclidean ring, we have

$$d\ [(ab)\ b^{-1}] \geq d\ (ab)$$

$$\therefore \quad d\ (a) \geq (ab). \qquad \ldots(2)$$

From (1) and (2), we conclude that

$$d\ (ab) = d\ (a).$$

(ii) Suppose now that b is not a unit in R. Since a and b are non-zero elements of the Euclidean ring R, therefore ab is also a non-zero element of R. Now $a \in R$ and $0 \neq ab \in R$, therefore by definition of Euclidean ring there exist elements q and in R such that

$$a = q\,(ab) + r \qquad \text{...(3)}$$

where either $r = 0$ or $d\,(r) < d\,(ab)$.

If $\quad r = 0$, then

$$a = qab$$

$\Rightarrow \quad a - qab = 0 \Rightarrow a\,(1 - qb) = 0$

$\Rightarrow \quad 1 - qb = 0 \qquad [\because a \neq 0$ and R is free of zero divisors]

$\Rightarrow \quad qb = 1 \Rightarrow$ b is inversible

$\Rightarrow \quad$ b is a unit in R.

Thus we get a contradiction. Hence r cannot be zero. Therefore we must have

$$d\,(r) < d\,(ab)$$

i.e., $\quad d\,(ab) > d\,(r)$. ...(4)

Also from (3), we have $r = a - qab = a\,(1 - qb)$.

$\therefore \quad d\,(r) = d\,[a\,(1 - qb)]$.

But $\quad d\,[a\,(1 - qb)] \geq d\,(a)$.

$\therefore \quad d\,(r) \geq d\,(a)$. ...(5)

From (4) and (5), we conclude that $d\,(ab) > d\,(a)$.

Theorem 8:

Let R be a Euclidean ring. Then every non-zero element in R is either a unit in R or can be written as a product of a finite number of prime elements of R.

Proof:

Let a be a non-zero element of R. We are to prove that either a is a unit in R or it can be written as a product of a finite number of prime elements of R. We shall prove that result by induction on d (a) *i.e.*, by induction on the d-value of a.

Let us first start the induction. We have a = 1a. Therefore $d\,(a) \geq d\,(1)$. Thus 1 is an element in R which has the minimal d-value. If $d\,(a) = d\,(1)$, then a is a unit in R. Thus the result of the theorem is true if $d\,(a) = d\,(1)$ and so we have started the induction.

Now assume as our induction hypothesis that the theorem is true for all non-zero elements $x \in R$ such that $d(x) < d(a)$. Then we shall show that the theorem is true for a also. If a is a prime element of R, the theorem is obviously true. So suppose that a is not prime. Then we can write $a = bc$ where neither b nor c is a unit in R. Since both b and c are not units in R, therefore $d(bc) > d(b)$ and $d(bc) > d(c)$. But $d(a) = d(bc)$. Therefore we have $d(b) < d(a)$ and $d(c) < d(a)$. So by our induction hypothesis each of b and c can be written as a product of a finite number of prime elements of R.

Let $b = p_1p_2 \ldots p_n$, $c = q_1q_2 \ldots q_m$ where the p's and q's are prime elements of R. Then

$$a = bc = p = p_1p_2 \ldots p_nq_1q_2 \ldots q_m.$$

Thus, we have written a as a product of a finite number of prime elements of R. This completes the induction and so the theorem has been proved.

Theorem 9:

Let a, b and c be any elements of a Euclidean ring R. Let $(a, b) = 1$ i.e., the greatest common divisor of a and b be 1. If $a \mid bc$, then $a \mid c$.

Proof:

If the greatest common divisor of a and b is 1, then by theorem 3 there exist elements λ and μ in R such that

$$1 = \lambda a + \mu b. \qquad \ldots(1)$$

Multiplying both members of (1) by c, we get

$$c = \lambda ac + \mu bc. \qquad \ldots(2)$$

But $a \mid bc$, so there exists an element $q \in R$ such that

$$bc = qa.$$

Substituting this value of bc in (2), we get

$$c = \lambda ac + \mu qa = (\lambda c + \mu q)\, a,$$

which shows that a is a divisor of c. Hence the theorem.

Example 1:

Let a be a given fixed real number and **R** *be the ring of real numbers Let ϕ be the mapping which associates with every polynomial $p(x)$ – with real coefficients–the real number $p(a)$. Show that ϕ is a homomorphism of the ring* **R** *[x] onto the ring* **R**.

Solution:

Here **R** is the ring of real numbers and **R** [x] is the ring of polynomials over the ring **R**.

We have **R** [x] = $\{a_0 + a_1x + ... a_nx^n : a_0, a_1, ..., a_n \in$ **R** and n is any integer $\geq 0\}$.

If $p(x) = a_0 + a_1x + ... + a_nx^n \in$ **R** [x], then the mapping ϕ: **R** [x] $\rightarrow$ **R** has been defined as

$$\phi(p(x)) = p(a) = a_0 + a_1a + ... + a_na^n.$$

To show that the mapping ϕ is a homomorphism of the ring **R** [x] onto the ring **R**.

The mapping ϕ is onto. Let $a \in$ **R**. Then $f(x) = a \in$ **R** [x] and by the definition of the mapping ϕ, we have

$$\phi(f(x)) = f(a) = a.$$

Thus $a \in$ **R** $\Rightarrow \exists\ f(x) = a \in$ **R** [x] such that $\phi(f(x)) = f(a) = a$.

Therefore the mapping ϕ is onto **R**.

Now to show that the mapping ϕ is a homomorphism.

ϕ Preserves Additions in R [x] and R

Let $f(x) = a_0 + a_1x + ... + a_nx^n$, $g(x) = b_0 + b_1x + ... + b_mx^m$ be any two elements of R [x]. Without loss of generality let $n \geq m$.

First let us take the case when $n > m$.

We have $\phi[f(x) + g(x)]$

$= \phi[(a_0 + b_0) + (a_1 + b_1)x + ... + (a_m + b_m)x^m + a_{m+1}x^{m+1} + ... + a_nx^n]$,

by definition of addition of two polynomials

$= (a_0 + b_0) + (a_1 + b_1)a + ... + (a_m + b_m)a^m + a_{m+1}a^{m+1} + ... + a_na^n$,

by definition of the mapping ϕ

$= (a_0 + a_1a + ... + a_na^n) + (b_0 + b_1a + ... + b_ma^m)$

$= f(a) + g(a)$

$= \phi[f(x)] + f[g(x)]$.

Similarly when $n = m$, we can show that

$$\phi[f(x) + g(x)] = \phi[f(x) + \phi[g(x)].$$

Thus $\phi[f(x) + g(x)] = \phi[f(x)] + \phi[g(x)],\ \forall\ f(x), g(x) \in$ R [x].

ϕ Preserves Multiplications is R [x] and R

We have $\phi[f(x)\, g(x)]$

$= \phi\,[a_0b_0 + (a_0b_1 + a_1b_0)\,a + (a_0b_2 + a_1b_1 + a_2b_0)\,x^2 + \ldots + a_nb_ma^{n+m},$

by definition multiplication of two polynomials

$= a_0b_0 + (a_0b_1 + a_1b_0)\,a + (a_0b_2 + a_1b_1 + a_2b_0)\,a^2 + \ldots + a_nb_ma^{n+m},$

by definition of the mapping ϕ

$= (a_0 + a_1 + \ldots + a_na^n)\,(b_0 + b_1a + \ldots + b_ma^m)$

$= f(a)\,g(a)$

$= \phi\,[f(x)]\,\phi\,[g(x)].$

Thus $\phi\,[f(x)\,g(x)] = \phi\,[f(x)\,\phi\,[g(x)],\ \forall\ f(x), g(x) \in \mathbf{R}\,[x].$

Hence the mapping ϕ is a homomorphism of the ring **R** [x] onto the ring **R**.

Example 2:

(a) Show that the set R of all elements of the form m + in, m, n ∈ Z, form a ring for addition and multiplication of complex numbers. Here Z is the ring of integers.

(b) Is the map q: R → Z, q (m + in) = m, a ring homomorphism?

(c) Give an example of a proper ideal in R.

Solution:

Here Z is the ring of integers and

$R = \{m + in,: m, n \in Z\}$ *i.e.*, R is the set of Gaussian integers.

To show that R is a ring for addition and multiplication of complex numbers.

Let a + ib and c + id be any two elements of R, where a, b, c, d ∈ Z.

Then $(a + ib) + (c + id) = (a + c) + i\,(b + d)$

and $(a + ib)\,(c + id) = (ac - bd) + i\,(ad + bc).$

These are again elements of R because a + c, b + d, ac – bd and ad + bc are all integers. Therefore R is closed with respect to ordinary addition and multiplication of complex numbers.

Further in complex numbers both addition and multiplication are associative as well as commutative compositions. Also multiplication distributes with respect to addition. The Gaussian integer 0 + i0 is the additive identity. If a + ib ∈ R, then (– a) + i (– b) *i.e.*, – a – ib is also an element of R and it is the additive inverse of a + ib because

$[(-a) + i\,(-b)] + (a + ib) = (-a + a) + i\,(-b + b) = 0 + i0$

$= \text{additive identity}.$

Hence R is a ring for addition and multiplication of complex numbers.

(b) Let θ: R $\to$ Z be defined as

$$\theta\ (m + in) = m,\ \forall\ m + in \in R.$$

Let u = a + ib and v = c + id be any two elements of R. Then

$$\theta\ (u + v) = \theta\ [(\theta + ib) + (c + id)] = \theta\ [(a + c) + i\ (b + d)]$$

$$= a + c, \text{ by definition of the mapping } \theta$$

$$= \theta\ (a + ib) + \theta\ (c + id) = \theta\ (u) + \theta\ (v).$$

Again $\theta\ (uv) = \theta\ [(a + ib)\ (c + id)] = \theta\ [(ac - bd) + i\ (ad + bc)]$

$$= ac - bd$$

while $\quad \theta\ (u)\ \theta\ (v) = \theta\ (a + ib)\ \theta\ (c + id) = ac.$

We observe that θ (uv) is not necessarily equal to θ (u) θ (v).

For example if u = 1 + 2i and v = 2 + 3i, then uv = (2 – 6) + (3 + 4) i = – 4 + 7i so that θ (u v) = – 4 while θ (u) θ (v) = 1.2 = 2. Thus,

$$\theta\ (uv) \neq \theta\ (u)\ \theta\ (v).$$

Thus the mapping θ preserves additions in R and Z but it does not preserve multiplications in R and Z.

We have $\theta\ (u + v) = \theta\ (u) + \theta\ (v),\ \forall\ u, v \in R$ but there exist elements u, v in R such that $\theta\ (uv) \neq \theta\ (u)\ \theta\ (v)$.

Hence the mapping θ is not a homomorphism of the ring R into the ring Z.

(c) *An example of a proper ideal i R.* To only units of the ring of Gaussian integers R are ± 1 and ± i.

Take a non-zero element of R which is not a unit of R. For example take u = 2 + 3i $\in$ R.

Let S = {zu: z $\in$ R}.

If z_1 u, z_2u are any two elements of S, then

$z_1u - z_2u = (z_1 - z_2)\ u \in S$ because $z_1 - z_2 \in R$.

Also if zu is any element of S and v is any element of R, then

v (zu) = (vz) u $\in$ S because vz $\in$ R

and also (zu) v $\in$ S because v (zu) = (zu) v.

$\therefore$ S is an ideal of R.

Obviously S is not the null ideal because

$1 = 1 + i0 \in R \Rightarrow 1\ (2 + 3i) = 2 + 3i \in S$ and so S contains some non-zero elements of R. Hence S is not the null ideal.

Also $S \neq R$ because $1 \in R$ but $1 \notin S$ as shown below.

If 1 is to be an element of S, then there must exist some $z \in R$ such that

$$zu = 1$$

$\Rightarrow$ $z(2 + 3i) = 1$

$\Rightarrow$ the complex number z is the multiplicative inverse of the complex number $2 + 3i$

$$\Rightarrow \quad z = \frac{1}{2+3i} = \frac{2-31}{(2+3i)(2-3i)} = \frac{2-3i}{13} = \frac{2}{13} - \frac{3}{13}i$$

which is not an element of R because 2/13 are not integers.

Thus there exists no $z \in R$ such that $zu = 1$.

$\therefore$ $1 \notin S$.

Thus $1 \in R$ while $1 \notin S$.

$S \neq R$.

Thus S is an ideal of R and S is neither the null ideal nor $S = R$.

Hence S is a proper ideal of R.

Example 3:

If U is an ideal of a ring R, then prove that R/U is a ring and is a homomorphic image of R.

Solution:

We have $R/U = \{U + a : a \in R\}$.

Here $U + a = \{u + a : u \in U\}$ and is called residue class of U in R generated by a

The operations of addition and multiplication in R/U are defined as follows:

$(U + a) + (U + b) = U + (a + b)$ [Addition of residue classes]

$(U + a)(U + b) = U + ab$ [Multiplication of residue classes]

To prove that R/U is a ring for these two compositions see the proof of theorem of this chapter.

Now to show that the ring R/U is a homomorphic image of the ring.

Consider the mapping $f: R \rightarrow R/U$ defined as

$$f(a) = U + a \ \forall \ a \in R.$$

The mapping f is onto. Let U + x be any element of R/U. Then $x \in R$.

We have f (x) = U + x. Therefore the mapping f is onto R/U.

The mapping f is a homomorphism. Let a, b $\in$ R. Then

$$f(a+b) = U + (a+b) = (U+a) + (U+b) = f(a) + f(b).$$

Also $f(ab) = U + ab = (U+a)(U+b) = f(a)\, f(b)$.

$\therefore$ f is a homomorphism of R onto R/U.

Hence the ring R/U is a homomorphic image of the ring R under.

The mapping f.

Example 4:

Let R be the ring of all real-valued continuous functions on the closed interval [0, 1] of the real line, the compositions being the usual pointwise addition and multiplication of functions. Let M be a subset of R defined by:

$$M = \{f \in R : f\left(\frac{1}{2}\right) = 0\}.$$

Prove that M is a maximal ideal of R.

Solution:

First of all we observe that M is non-empty because the real valued function e (x) on [0, 1] defined by

$$e(x) = 0 \ \forall \ x \in [0, 1]$$

belongs to M.

Now let f (x), g (x) be any two elements of M. Then

$$f\left(\frac{1}{2}\right) = 0,\ g\left(\frac{1}{2}\right) = 0, \text{ by definition of M.}$$

Let $\quad h(x) = f(x) - g(x)$. Then

$$h\left(\frac{1}{2}\right) = f\left(\frac{1}{2}\right) - g\left(\frac{1}{2}\right) = - 0 = 0.$$

Therefore $h(x) \in M$.

Thus $\quad f(x), g(x) \in M$

$\Rightarrow \quad h(x) = f(x) - g(x) \in M$.

Further let f (x) be any element of M and r (x) be any element of R.

Then $f\left(\frac{1}{2}\right) = 0$, by definition of M.

Let t (x) = r (x) f (x) = f (x) r (x). [∴ R is a commutative ring]

Then $t\left(\frac{1}{2}\right) = r\left(\frac{1}{2}\right) f\left(\frac{1}{2}\right) = r\left(\frac{1}{2}\right) . 0 = 0$. Therefore (x) Î M

Thus r (x) ∈ R, f (x) ∈ M ⇒ r (x) f (x) ∈ M.

Hence M is an ideal of R.

Clearly M ∈ R because i (x) ∈ R defined by i (x) = 1 ∀ x ∈ [0, 1] does not belong to M.

The ring R is with unity and the element i (x) is its unity element.

Let N be an ideal of R properly containing M *i.e.*, M ⊆ N and M ≠ N. Then M will be a maximal ideal of R if N = R, which will be so if the unity i (x) of R belongs to N. Since M is a proper subset of N, therefore there exists λ (x) ∈ N such that λ (z) ∉ M. This means $\lambda\left(\frac{1}{2}\right) \neq 0$. Put $\lambda\left(\frac{1}{2}\right) = c$ where c ≠ 0.

Let us define β (x) ∈ R by β (x) = c ∀ x ∈ [0, 1]. Now consider μ (x) ∈ R given by μ (x) = λ (x) – β (x).

We have $m\left(\frac{1}{2}\right) = \lambda\left(\frac{1}{2}\right) - \beta\left(\frac{1}{2}\right)$.

Therefore μ (x) ∈ M and so μ (x) also belongs to N because N is a superset of M. Now N is an ideal of R and λ (x), μ (x) are in N. Therefore λ (x) - μ (x) = β (x) is also an element of N.

Now define γ (x) ∈ R by γ (x) = 1/c ∀ x ∈ [0, 1]. Since N is an ideal of R, therefore γ (x) ∈ R and β (x) ∈ N ⇒ γ (x) β (x) ⊂ N.

We shall show that γ (x) β (x) = i (x).

For every x ∈ [0, 1], we have

γ (x) β (x) = (1/c) = 1.

Therefore γ (x) β (x) = i (x), by definition of i (x).

Thus the unity element i (x) of R belongs to N and consequently N = R.

Hence M is a maximal ideal of R.

Example 5:

If U is a left ideal of a ring R, let

$\lambda(U) = \{x \in R: xu = 0 \ \forall u \in U\}$.

Prove that $\lambda(U)$ is a two sided of R.

Solution:

First we see that $\lambda(U) \neq \varnothing$ because $0 \in R$ is such that

$$0u = 0 \ \forall \ u \in U.$$

Now let x_1, x_2 be any two elements of $\lambda(U)$. Then

$x_1 u = 0 \ \forall \ u \in U$ and $x_2 u = 0 \ \forall \ u \in U$.

We have $(x_1 - x_2) u = x_1 u - x_2 u = 0 - 0 = 0$ for all $u \in U$.

$\therefore \quad x_1 - x_2 \in \lambda(U)$.

Now let x be any element of $\lambda(U)$ and r be any element of R. Then

$xu = \ \forall \ u \in U$ [by def. of $\lambda(U)$

$\Rightarrow \ r(xu) = r0 \ \forall \ u \in U$

$\Rightarrow \ (rx)u =$ for all $u \in U \Rightarrow rx \in \lambda(U)$.

Further U is a left ideal of R. Therefore $ru \in U \ \forall \ u \in U$.

Since $x \in \lambda(U)$, therefore by def. of $\lambda(U)$, we have

$x \in \lambda(U)$, $ru \in U \Rightarrow x(ru) = 0$ for all $u \in U$

$\Rightarrow (xr)u = 0$ for all $u \in U$

$\Rightarrow xr \in \lambda(U)$.

Thus $x \in \lambda(U)$, $r \in R \Rightarrow xr, rx \in \lambda(U)$

Hence $\lambda(U)$ is a two sided ideal of R.

Example 6:

Show that an arbitrary intersection of ideals of a ring is an ideal of the ring.

Solution:

Let R be a ring and let $\{S_t : t \in T\}$ be any family of ideals of R. Here T is an index set and is such that $\forall \ t \in T$, S_t is an ideal of R.

Let $\quad S = \bigcap\limits_{t \in T} S_t = \{x \in R: x \in S_t \ \forall \ t \in T\}$

be the intersection of this family of ideals of R. Then to prove that S is also an ideal of R.

Obviously $S \neq \emptyset$, since at least 0 is in $S_t\ \forall\ t \in T$

Now let a, b be any two elements of S. Then

$a, b, \in S \Rightarrow a, b \in S_t\ \forall\ t \in T$

$\Rightarrow a - b\ t \in S_t\ \forall\ t \in T$

[$\because$ $\forall\ t \in T$, S_t is an ideal of R]

$\Rightarrow a - b \in \bigcap_{t \in T} S_t \Rightarrow a - b \in S.$

$\therefore$ S is a subgroup of the additive group of R.

Now let s be any element of S and r be any element of R.

We have $s \in S \Rightarrow s \in \bigcap_{t \in T} S_t$

$\Rightarrow s \in S_t\ \forall\ t \in T$

$\Rightarrow rs \in S_t$ and $sr \in S_t\ \forall\ t \in T$

[$\because$ $\forall\ t \in T$, S_t is an ideal of R]

$\Rightarrow rs \in \bigcap_{t \in T} S_t$ and $sr \in \bigcap_{t \in T} S_t$

$\Rightarrow rs \in S$ and $sr \in S$.

Thus $a, b, \in S \Rightarrow a - b \in S$

and $r \in R, s \in S \Rightarrow rs \in S, sr \in S$.

Hence S is an ideal of R.

Example 7:

If U, V are ideals of a ring R let UV be the set of all those elements of R which can be written as finite sums of elements of the form uv where $u \in U$ and $v \in V$. Prove that UV is an ideal of R.

Also show that $UV \subseteq U \cap V$

Solution:

U and V are ideals of a ring R, Let

$UV = \{u_1v_1 + u_2v_2 + \ldots + u_nv_n : u_1, u_2, \ldots, u_n \in U, v_1\ v_2, \ldots, v_n \in V$ and n is any positive integer$\}$.

To prove the UV is also an ideal of R.

Let $\alpha\ u_1v_1 + u_2v_2 + \ldots + u_nv_n$, $\beta = u_1'v_1' + u_2'\ v_2' + \ldots + u_m'v_m'$ be and two elements of UV, where $u_1, u_2, \ldots, u_n, u_1, u_2, \ldots, u_m' \in U$ and $v_1, v_2, \ldots, v_n, v_1', v_2', \ldots, v_m' \in V$. Also m and n are any positive integers.

We have $\alpha - \beta = u_1v_1 + u_2v_2 + ... + u_nv_n - u_1'v_1' - u_2'v'_m = u_1v_1 + u_2v_2 + ... + u_nv_n + (-u_1')\, v_1' + (-u_2')\, v_2' + ... + (-u_m')\, v_m'$.

This is obviously an element of UV because U is an ideal and therefore $u_1' \in U \Rightarrow (-u_1') \in U$, etc.

Again let $r \in R$ and $\alpha \in UV$. Then

$r\alpha = r\,(u_1v_1 + u_2v_2 + ... + u_nv_n) = (ru_1)\, v_1 + (ru_2)\, v_2 + ... + (ru_n)\, v_n$.

This is an element of UV because U is an ideal and therefore $r \in R$, $u_1 \in U \Rightarrow ru_1 \in U$, etc.

Also $\alpha r = (u_1v_1 + u_2v_2 + ... + u_nv_n)\, r = u_1\,(v_1r) + u_2\,(v_2r) + ... + u_n\,(v_nr)$.

This is an element of UV because V is an ideal and therefore $r \in R$, $v_1 \in V \Rightarrow v_1r \in V$, etc.

Hence UV is an ideal of R.

Now to show that $UV \subseteq U \cap V$.

Let $\alpha = u_1v_1 + ... + u_nv_n$ be any element of UV where

$$u_1, ... u_n \in U, \text{ and } v_1, ..., v_n \in V.$$

Now $v_1 \in V \Rightarrow v_1 \in R$. Also U is an ideal. Therefore

$$v_1 \in R,\ u_1 \in U \Rightarrow u_1v_1 \in U.$$

Similarly $u_1 \in U \Rightarrow u_1 \in R$. But V is an ideal. Therefore

$$u_1 \in R,\ v_1 \in V \Rightarrow u_1v_1 \in V.$$

Thus $\quad u_1v_1 \in U,\ u_1v_1 \in V \Rightarrow u_1v_1 \in U \cap V$.

Similarly $u_2\, v_2, ..., u_nv_n \in U \cap V$.

Since $\quad U \cap V$ is also an ideal of R, therefore

$$u_1v_1, ..., u_nv_n \in U \cap V \Rightarrow \alpha = u_1v_1 + ... + \in U \cap V.$$

Thus $\alpha \in UV \Rightarrow \alpha \in U \cap V$. Therefore $UV \subseteq U \cap V$.

Example 8:

For any given element a of a ring R let

$$Ra = \{\, xa : x \in R \}.$$

Prove that Ra is a left ideal of R.

Solution:

Let x_1a, x_2a be any two elements of R a where $x_1, x_2 \in R$. We have $x_1a - x_2a = (x_1 - x_2)\, a \in R$ Ra, since

$$x_1, x_2 \in R \Rightarrow x_1 - x_2 \in R.$$

Thus $x_1a, x_2a \in Ra \Rightarrow x_1a - x_2a \in Ra$.

Now let xa be any element of Ra where $x \in R$ and r be any element of R. We have $r(xa) = (rx)a \in Ra$, since

$$r \in R, x \in R \Rightarrow rx \in R.$$

Thus $r \in R$, $xa \in Ra \Rightarrow r(xa) \in Ra$.

$\therefore$ ra is a left ideal of R.

Example 9:

If U is an ideal of a ring R, let

$$[R: U] = \{x \in R: rx \in U \ \forall \ r \in R\}.$$

Prove that [R: U] is an ideal of R and that it contains U.

Solution:

First we see that [R: U] is not empty because $0 \in R$ is such that $r0 = 0 \in U$ for all $r \in R$.

Now let x_1, x_2 be any two elements of [R: U]. Then

$rx_1 \in U \ \forall \ r \in R$, and $rx_2 \in U \ \forall \ r \in R$,

Since U is an ideal, therefore

$$rx_1 \in U, rx_2 \in U \Rightarrow rx_1 - rx_2 \in U$$
$$\Rightarrow r(x_1 - x_2) \in U \ \forall \ r \in R$$
$$\Rightarrow x_1 - x_2 \in [R: U], \text{ by def. of } [R: U].$$

Now let x be any element of [R: U] and s be any element of R. Then $rx \in U \ \forall \ r \in R$

$\Rightarrow (rx)s \in U \ \forall \ r \in R$ $\quad$ [$\because$ U is an ideal and so $s \in R$, $rx \in U \Rightarrow (rx)s \in U$]

$\Rightarrow r(xs) \in U$ for all $r \in R \Rightarrow xs \in [R: U]$.

Also $rx \in U \ \forall \ r \in R \Rightarrow sx \in U$ $\quad$ [$\because$ $s \in R$]

$$\Rightarrow (sx)r \in U \ \forall \ r \in R$$

[$\because$ U is an ideal and so $sx \in U$, $r \in R \Rightarrow (sx)r \in U$]

$\Rightarrow sx \in [R: U]$.

Thus $x \in [R: U]$, $s \in R \Rightarrow xs \in [R: U]$, $sx \in [R: U]$.

Therefore [R: U] is an ideal or R.

Now to show that $U \subseteq [R: U]$. We have

$y \in U \Rightarrow yr \in U \ \forall \ r \in R$ $\quad$ [$\because$ U is an ideal]

$\Rightarrow y \in [R: U]$.

$\therefore\ U \subseteq [R: U]$.

Theorem 1:

A commutative ring with unity is a field if it has no proper ideals.

Proof:

Let r be a commutative ring with unity having no proper ideals *i.e.*, the only ideals of R are (0) and R itself. In error to show that R is aa field, we should show that each non-zero element of R possesses multiplicative inverse.

Let a be any non-zero element of R.

The set $Ra = \{ra : r \in R\}$ is an ideal of R.

Since $1 \in R$, therefore $1a = a \in Ra$. Thus, $0 \neq a \in Ra$. Therefore the ideal $Ra \neq (0)$. Since R has no proper ideals, therefore the only possibility is that Ra = R. Thus every element of R is a multiple of a by some element of R. In particular, $1 \in R$ so its can be realised as a multiple of a. Thus there exists an element $b \in R$ such that ba = 1. Therefore $a^{-1} = b$. Hence each non-zero element of R possesses multiplicative inverse.

$\therefore$ R is field.

Theorem 2:

A field has no proper ideals i.e., if F is a field then its only ideals are (0) and F itself.

Proof:

Let S be any non-zero ideal of the field F and let a be any non-zero element of S. We have $a^{-1} \in F$.

Since S is an ideal, therefore

$a \in S,\ a^{-1} \in F \Rightarrow aa^{-1} \in S \Rightarrow 1 \in S$.

Now let x be any element of F. Then

$1 \in S,\ x \in F \Rightarrow 1x \in S \Rightarrow x \in S$.

Thus each element of F belongs to S. Therefore $F \subseteq S$. But $S \subseteq F$. Therefore $S \subseteq F$.

Thus the only ideals of F are (0) and F itself.

Theorem 3:

If R is a commutative ring and $a \in R$, then

$Ra = \{ra: r \in R\}$ is an ideal of R.

Proof:

In order to prove that Ra is an ideal of R, we should prove that Ra is a subgroup of R under addition and that if $u \in Ra$ and $x \in R$ then xu and ux are also in Ra. But R is a commutative ring, therefore xu = ux. Thus we only need to check that xu is in Ra.

Now, let u, v, $\in$ Ra. Then $u = r_1a$, $v = r_2a$ for some $r_1, r_2 \in R$.

We have $u - v = r_1a - r_2a = (r_1 - r_2)\, a \in Ra$ since $r_1 - r_2 \in R$.

Thus $u, v \in Ra \Rightarrow u - v \in Ra$. Hence Ra is a subgroup of R under addition.

Now let $x \in R$.

Then $xu = x\,(r_1a) = (xr_1)\, a \in Ra$ since $xr_1 \in R$.

$\therefore$ Ra is an ideal of R.

3.15 IDEAL GENERATED BY A GIVEN SUBSET OF A RING

If M is any subset of a ring R, we can find ideals containing M. For example the ring r itself is an ideal containing any subset of R.

Smallest Ideal Containing a Subset

Let M be any arbitrary subset of a ring R. Then an ideal S of R is called the smallest ideal of R Containing M if

$$M \subseteq S.$$

and if S is contained in every ideal of R containing M.

Definition: *Let R be a ring and let M be any arbitrary subset of R. The smallest ideal of R containing M is said to be the ideal generated by M and is denoted by (M).*

In particular if M consists of single element, say a, of the going R we write (a) in place of (M). An ideal such as (a) generated by a single element of the ring is called a *principal ideal.*

Principal Ideal

Definition: *An ideal S of a ring R is said to be a principal ideal if there exists an element $a \in S$ such that an ideal To of R containing a also contains S i.e., $S = (a)$.*

Thus, an ideal generated by a single element of itself is called a principal ideal.

Theorem 1:

If a is an element in a commutative ring R with unity, then the set S = {ra: r ∈ R} is a principal ideal of R generated by the element a i.e., S = (a).

Proof:

First we should prove that a ∈ S. Since R is a ring with unit element 1, therefore 1a = a ∈ S.

Now we should prove that S is an ideal of R. So first we should prove that S is a subgroup of R under addition. Let u, v be any two elements of S. Then u = r_1a, v = r_2a for some r_1, r_2 ∈ R.

We have u – v $r_1a - r_2a = (r_1 - r_2)$ a ∈ S since $r_1 - r_2$ ∈ R.

∴ S is a subgroup of R under addition.

Now we should prove that x ∈ R, u ∈ S ⇒ xu ∈ S and ux ∈ S. But R is commutative ring, therefore xu = ux and thus it is sufficient to show that xu ∈ S.

We have xu = x (r_1a) = (xr_1) a ∈ S since xr_1 ∈ R.

∴ S is an ideal of R and a ∈ S.

Now under to prove that S is an ideal generated by the element a, we should prove that if T is an ideal of R and a ∈ T, then S ⊆ T.

Let ra be any element of S. Then r ∈ R. If T is an ideal of R containing a, then a ∈ T, r ∈ R ⇒ ra ∈ T. Thus S ⊆ T.

Hence S is a principal of R generated by the element a.

Theorem 2:

Let S be an ideal of a commutative ring R. Let a be an element of S such that

x ∈ S ⇒ x = ya for some y ∈ R.

Then S is a principal of R generated by a.

Proof:

As given in the statement of the theorem, S is an ideal or R containing the element a. Let T be any ideal of R containing a. Then s will be a principal ideal of R generated by a if S ⊆ T.

Let x be any element of S. Then x = ya for some y ∈ R.

Now y ∈ R, a ∈ T ⇒ ya ∈ T [∵ T is an ideal]

$\Rightarrow \; x \in T.$ [$\because$ $x = ya$]

Thus $\quad x \in S \Rightarrow x \in T.$

$\therefore S \subseteq T.$

Hence S is a principal ideal of R generated by a.

Example:

Suppose we are to find the principal ideal generated by 5 in the ring of integers. The ring I of integers is a commutative ring with unity. Therefore (5) = {5r: r ∈ I}.

Thus the principal ideal generated by 5 is given by

(5) = {..., – 10, –5, 0, 10,...}.

Obviously (–5) = (5).

3.16 PRINCIPAL IDEAL RING

Definition: *A commutative ring R without zero divisors and with unity element is a principal ideal ring if every ideal S in R is a principal ideal i.e., if every ideal S in R is of the form S = (a) for some a ∈ S.*

Theorem 1:

Every field is a principal ideal ring.

Proof:

A field has no proper ideals. Then only ideals of a field are (1) the null ideal which is a principal ideal generated by 0 and (2) the field itself which is also a principal ideal generated by

1. Thus a field is always a principal ideal ring

Theorem 2:

The ring of integers is a principal; ideal ring.

Proof:

Let (I, +, ·) be the ring of integers. Obviously I is a commutative ring with unity and without zero divisors. Therefore I will be a principal ideal ring if every ideal in I is a principal ideal.

Let S be any ideal of the ring of integers. If S is the null ideal then S = (0) so that S is a principal ideal.

So let us suppose that S ≠ (0).

Now S contains at least one non-zero integer, say a. Since S is a subgroup of R under addition, therefore $a \in S \Rightarrow -a \in S$. This shows that S contains at least one positive integer, because if $0 \neq a$, then one of a and $-a$ must be positive.

Let S_+ be the set of all positive integers in S. Since S_+ is not empty, therefore by the well ordering principle S_+ must possess a lest positive integer. Let s be this least element. We will now show that S is the principal ideal generated by s *i.e.*, S = (s).

Suppose now that is integer in S. Then by division algorithm, there exist integers qu and r such that $n = qs + r$ with $0 \leq r < s$.

Now $\quad s \in S, q \in I \Rightarrow qs \in S$ $\quad$ [$\because$ S is an ideal]

and $\quad n \in S, qs \in S \Rightarrow n - qs \in S$ $\quad$ [$\because$ S is a subgroup of the additive group of I]

$\Rightarrow r \in S.$ $\quad$ [$\because n - qs = r$]

But $0 \leq r < s$ and s is the last positive integer such that $s \in S$. Hence r must be 0.

$\therefore \quad n = qs.$

Thus $n \in S \Rightarrow n = qs$ for some $q \in I$.

Hence S is a principal ideal of I generated by s.

Since S was an arbitrary ideal in the ring of integers, therefore the ring of integers is a principal ideal ring.

3.17 DIVISIBILITY IN AN INTEGRAL DOMAIN

Definition: *Suppose $0 \neq a$ is an element of a commutative ring R. Then a is said to divide $b \in R$,* if there exists an element $c \in$ *R such that b = ca.*

We shall use the symbol $a \mid b$ to represent the fact that a divides b. Also if a divides b then sometimes we say that a is a factor of b or b is divisibility it follows that every non-zero element of r is divisor of its zero element. Obviously we can write $0 = a0$. Therefore if $0 \neq a \in R$, then $a \mid 0$.

Theorem:

If R is a commutative ring, then

1. *$a \mid b$ and $b \mid c \Rightarrow a \mid c$ i.e., the relation of divisibility in R is a transitive relation.*
2. *$a \mid b$ and $a \mid c \Rightarrow a \mid (b + c)$*
3. *$a \mid b \Rightarrow a \mid bx$ for all $x \in R$.*

Proof:

1. $a \mid b \Rightarrow b = ap$ for some $p \in R$

and $b \mid c \Rightarrow c = aq$ for some $q \in R$.

Now $c = bq$ and $b = ap \Rightarrow c = (ap)\, q \Rightarrow c = a(pq)$

$\Rightarrow a \mid c$ since $pq \in R$.

2. $a \mid b \Rightarrow b = ap$ for some $p \in R$

and $a \mid c \Rightarrow c = ap$ for some $q \in R$

Now $b = ap$ and $c = aq \Rightarrow b + (ap + aq \Rightarrow b + c\ a\ (p + q)$

$\Rightarrow a \mid (b + c) =$ since $(p + q) \in R$.

3. $a \mid b \Rightarrow b = aq$ for some $p \in R$.

Now $b = ap \Rightarrow bx = (ap)\, x\ \forall\ x\ R$.

$\Rightarrow bx = a\,(px) \Rightarrow a \mid b$ since $px \in R$.

Example 1:

In the ring I of integers, we have $3 \mid 6$ since we have $6 = 3 \times 2$ and $2 \in R\ I$.

However in the ring of integers 3 is not a divisor of 7.

Solution:

Do yourself.

Example 2:

In the ring Q of rational numbers, we have $3 \mid 7$ since we have $7 = 3 \times (7/3)$ and $7/3 \in Q$.

Solution:

Do yourself.

Units: *Let R be a commutative ring with unity element 1. An element $a \in R$ is a unit in R if there exists on element $b \in R$ such that $ab = 1$.*

In the ring of integers I, the only units are 1 and –1. These are the only inversible elements of the ring of integers.

Every non-zero element of a field possesses multiplicative inverse. Therefore every non-zero element of a field is a unit.

It is obvious that if a is a unit in a ring R, then a^{-1} is also a unit in R. Also the product of two units is again a unit. Because if a, b are two units in R, then $(ab)^{-1} = b^{-1} a^{-1}$ which is an element or R. Of course the set of all units in R forms a group under multiplication.

Example:

Find all the units of the integral domain of Gaussian integers.

Solution:

Let D = {a + ib: a, b ∈ I the set of integers} be the ring of Gaussian integers. The element 1 + 0i is the unity element of this ring. Let x + iy be unit and x' + iy' be its inverse.

Then $(x + iy)(x' + iy') = 1 + 0i$

$\Rightarrow$ $(xx' - yy') + i(xy' - yx') = 1 + 0i.$

Equating real and imaginary parts, we get

$$xx' - yy' = 1 \quad ...(1)$$

and $$xy' + yx = 0. \quad ...(2)$$

Squaring and adding (1) and (2), we get

$$x^2x'^2 + y^2y'^2 + x^2y'^2 + y^2x'^2 = 1$$

$\Rightarrow$ $$(x^2 + y^2)(x'^2 + y'^2) = 1.$$

Now the product of two positive integers can be equal to 1 if and only if each of them is 1.

$\therefore$ $$x^2 + y^2 = 1.$$

This given $x^2 = 0, y^2 = 1$

$\Rightarrow$ $x^2 = 1, y^2 = 0.$

Thus $x = 0, y = \pm 1$

$\Rightarrow$ $x = \pm 1, y = 0.$

$\therefore$ The only units of the integral domain of Gaussian integers are 0 ± i, (±1) + 0i *i.e.*, 1, –1, i, –t.

Associates

Definition: *Let R be a commutative ring with unity element 1. Then an element a of R is said to be an associate of b ∈ R if a = ub for some unit u in R.*

In symbols, we express it by writing a ~ b which is read as 'a is an associate of b' or a and b are associates'

From this definition we observe that in a commutative ring with unity all the associates of an element can be obtained by multiplying that element by different units in that ring.

Illustrations

1. The only units of the integral domain of integers are 1 and –1. Therefore if a is any non-zero integer, then it has exactly two associates namely 1a and (–1) a *i.e.*, a and –a. Thus the two associates of 5 are 5 and –5.
2. In any commutative ring with unity the associate of 0 is only zero.
3. The only units of the domain of Gaussian integers are 1, –1, i, –i. Therefore if a + ib is any non-zero element of this domain, then it has exactly four associates namely,

$$1(a + ib),\ -1\ (a + ib),\ i\ (a + ib),\ -i\ (a + ib)$$

i.e. $a + ib,\ -a - ib,\ -b + ia,\ b - ia.$

Theorem 1:

Let D be an integral domain with unity element 1. Two non-zero elements a, b $\in$ D are associates if and only if a | b and b | a.

Proof:

Let two non-zero elements a, b be associates of each other in D. Then a = bu where u is a unit in D.

Now $a = bu \Rightarrow b \mid a$.

Again $a = bu \Rightarrow au^{-1} = b \cup u^{-1} \Rightarrow au^{-1} = b \Rightarrow b = au^{-1} \Rightarrow a \mid b$.

Thus a and b are associates $\Rightarrow a \mid b$ and $b \mid a$.

Conversely, let a | b and b | a. Then to prove that a and b are associates.]

We have $a \mid b \Rightarrow \exists\ c \in D$ such that b ac.

Similarly $b \mid a \Rightarrow \exists\ d \in D$ such that $a = bd$.

$\therefore b = ac = (bd)\ c = d\ (dc)$.

$\therefore b1 = b\ (dc)$ [$\because$ $b1 = b$]

$\Rightarrow b\ (1 - dc) = 0$

$\Rightarrow (1 - dc) = 0$ [$\because$ $b \neq 0$ and D is without zero divisors]

$\Rightarrow 1 = dc$.

$\therefore$ both c and d are units in D.

Thus a = bd where d is a unit in d. Hence a and b are associates.

Note: In a field any two non-zero elements are associates.

Theorem 2:

Let R be a commutative ring with unity element 1. The relation in R defined by 'a is an associate of b' is an equivalence relation.

Proof:

Reflexivity: Let a be any element of R. Then a = 1a. Therefore a ~ a because 1 is a unit in R. Thus ~ is reflexive.

Symmetry: We have $a \sim b \Rightarrow a = ub$ for some unit u in R $\Rightarrow u^{-1} a = u^{-1} ub \Rightarrow u^{-1}a = 1b\, u^{-1} a = b \Rightarrow b = u^{-1}a \Rightarrow b \sim a$ because u^{-1} is also a unit in R. Therefore ~ is symmetric.

Transitivity: Let a ~ b, b ~ c. Then a = ub and b = vc for some units u, v $\in$ R. This gives a = u (vc) = (uv) c. But the product of two units in R is again a unit in r. Therefore uv is a unit in R and so a (uv) c $\Rightarrow$ a ~ c. Therefore ~ is transitive.

Hence ~ is an equivalence relation in R.

3.18 PROPER AND IMPROPER DIVISORS

Definition: *Let D be an integral domain with unity element 1. Let a be any non-zero element of D. Then the units of D and associates of a are always divisors of a. These are called improper or trivial divisors of a. Any other divisors of a are called proper or non-trivial divisors of a.*

Prime Elements

Definition: *Let D be an integral domain with unity element 1. A non-zero non-unit element a $\in$ D, having only trivial divisors, is called a* **prime** or **irreducible** *element of D.* An element $0 \neq b \in D$ having proper divisors is called a *reducible* or *composite* element of D. From this definition it is obvious that if p is a prime element of D and if p = xy, where x, y $\in$ D, then one of x or y must be unity in D.

(Raj. 77)

Also $0 \neq b \in D$ is a composite element of D if and only if we can find two elements x, y $\in$ D such that b = xy and none of x and y is a unit in D.

Greatest common divisor

Definition: *Let R be a commutative ring. If a, b $\in$ R, then $0 \neq d \in R$ is said to be a greatest common divisor of a and b if*

1. d | a and d | b.
2. whenever c | a and c |b then c | d.

We shall use the notation d = (a, b) to denote that d is a greatest common divisor of a and b.

Now suppose a, b $\in$ D where D is an integral domain with unity element 1. Let a, b possess a greatest common divisor.

If d_1, d_2 are two greatest common divisors of a and b, we have

$$d_1 \mid d_2 \text{ and } d_2 + d_1$$

$$\Rightarrow d_1 \text{ and } d_2 \text{ are associates.}$$

Thus in an integral domain with unity in case a greatest common divisor of a and b exists, it is unique apart from the distinction between associates.

3.19 RELATIVELY PRIME ELEMENTS

Definition: *Let D be an integral domain with unity element 1. Two elements a, b $\in$ D are said to be relatively prime if their greatest common divisor is a unit of D.*

But any associate of a greatest common divisor is a greatest common divisor. Also the unity element 1 is an associate of any unit. Therefore if a, b are relatively prime we may assume that a greatest common divisor of a and b is 1 *i.e.*, (a, b) = 1.

Polynomial Rings

While studying algebra in high school classes we are introduced with polynomials. We know that expressions of the type $3x^2 - 4x\ 5$, $x + 7$, $9x^3 - \frac{2}{3}x^2 + 4x + 5$ etc, are called polynomials in the Indeterminate x. In place of x we can use other letters like y, z etc., to denote these polynomials. Now we shall define polynomials over an arbitrary ring.

Definition: *Let R be an arbitrary ring and let x, called an* **indeterminate,** *be any symbol not an element of R. By a polynomial in x over R is meant an expression of the form*

$$f(x) = a_0x^0 + a_1x + x_2x^2 + \dots \text{ where}$$

a_0, a_1, a_2,..., are elements of R and only a finite number of them are not equal to 0, the zero element or R.

Here x is an indeterminate. We could have used nay other letter, say, y in place of x. Also a_0x^0, a_1x, a_2x^2 etc., are called terms of the polynomial and a_0, a1, a2, etc., are called coefficients of these terms. All these coefficients are elements of R. The number of terms in the polynomial f (x) will be infinite but except a finite number of terms, the coefficients of all the remaining terms will be equal to 0, the zero element of the ring. The symbol '+' connecting various terms in f (x) has no connection with the addition of the ring. This symbol has been used here only to connect different terms. Also x is not an element of R. The powers of x are nothing to do with the powers of an element of R. The are nothing to do with the powers of an element of R. The different powers of x only tell us the ordered place of different coefficients. There is no harm if we represent this polynomial f (x) be the infinite ordered st $(a_0, a_1, a_2,...)$, where $a_0, a_1, a_2,...$ are elements of R and only a finite number of them are not equal to zero. Since from his school classes we represent a polynomial with an indeterminate x, therefore we have preferred this way to represent polynomials.

Remark:

The polynomial $a_0x^0+a_1x+a_2x^2+...$ *over a ring R can also be written as*

$$a_0 + a_1x + a_2x^2 + ...$$

Set of all polynomials over a ring. Let R be an arbitrary ring and x an indeterminate. The set of all polynomials f (x),

$$f(x) = \sum_{n-0}^{\infty} a_n x^n = a_0x^0 + a_1x + x_2x^2 + ...$$

where the a' s are elements of the ring R and only a finite number of then are not equal to zero, is called R [x].

We shall make a ring out of R [x]. Then R [x] will be called the ring of all polynomials over the ring R. For this we shall define *equality, addition and multiplication* of two elements of R [x].

Definition: *Suppose R is an arbitrary ring and*

$$f(x) = a_0x^0 + a_1x + a_2x^2 = a_2x^2 + ...$$

and $$g(x) = b_0x^0 + b_1x + b_2x^2 = b_2x^2 + ...$$

(a) $f(x) + g(x)$ *if and only if* $a_n = b_n \ \forall$ *non-negative integer n. Thus two polynomials are equal iff their corresponding coefficients are equal.*

(b) $f(x) + g(x) = = c_0x^0 + c_1x + c_2x^2 + c_3x^3 + ...$ *where* $c_n = a_n + b_n$ *fro every non-negative integer n. Thus in order to add two polynomials we should add the coefficients of like powers of x.*

Since $c_n \in R$ and only a finite number of c's can be not equal to zero, therefore f (x) + g(x) is an element of R [x]. Thus R [x] is closed with respect to addition of polynomials as defined above.

(c) $f(x)\, g(x) = d_0x^0 + d_1x + d_2x^2 + d_3x^3 + ...$

where $$d_n = a_0b_n + a_1b_{n-1} + a_2b_{n-2} + ... + a_nb_0$$

for every non-negative integer n. We can write $d_n = \sum_{i+j-n} a_i b_j$ where by this summation we mean the sum of all the products of the type $a_i b_j$ with i and j non-negative integers whose sum is n.

Since $d_n \in R$ and only a finite number of d's can be not equal to zero, therefore f (x) g (x) is an element of R [x]. Thus R [x] is closed with respect to multiplication of polynomials as defined above.

We have $d_0 = a_0 b_0$, $d_1 = a_0 b_1 + a_1 b_0$,

$d_2 = a_0 b_2 + a_1 b_1 + a_2 b_0$, $d_3 = a_0 b_3 + a_1 b_2 + a_2 b_1 + a_3 b_0$ and so on.

Therefore in order to multiply two polynomials f (x) and g (x), we should first write

$$f(x)\, g(x) - (a_0 x^0 + a_1 x + a_2 x^? + \ldots)(b_0 x^0 + b_1 x + b_2 x^2 + \ldots)$$

Now we should multiply different powers of the determinate x and using the relation $x^i x^j = x^{i+j}$ we should collect coefficients of different powers of x.

Zero Polynomial. The polynomial

$$f(x) = \sum a_n x^n = a_0 x^0 + a_1 x + a_2 x^2 + a_3 x^3 + \ldots$$

in which all the coefficients $a_0, a_1, a_2, \ldots$ are equal to 0 is called the zero polynomial over the ring R.

3.20 DEGREE OF A POLYNOMIAL

Let $f(x) = a_0 x^0 + a_1 x + a_2 x^2 + a_3 x^3 + \ldots + a_n x^n + \ldots$

be a polynomial over an arbitrary ring R. We say that n is the degree of the polynomial f (x) if and only if $a_n \neq 0$ *and* $a_m = 0$ *for all* $m > n$. We shall write deg f (x) is the largest non negative integer i for which the ith coefficient of f (x) is not 0. If in the polynomial f (x), a_0 (*i.e.*, the coefficient of x^0) is not 0 and all the other coefficients are 0, then according to our definition, the degree of f (x) will be zero. Also according to our definition, if there is no non-zero coefficient in f (x), then its degree will remain undefined. Then we do not define the degree of the zero polynomial. Also it is obvious that very non-zero polynomial will possess a unique degree.

Note: If $f(x) = a_0 x^0 + a_1 x + a_2 x^2 + \ldots + a_n x^n + \ldots$ is a polynomial of degree n *i.e.*, if $a_n \neq 0$ and $a_m - 0$ for all $m > n$, then it is convenient to write f (x) $= \sum_{i-0}^{n} a_i x^i = a_0 x^0 + a_1 x + a_2 x^2 + \ldots + a_n x^n$. It will remain understood that all the terms in f (x) which follow the term $a_n x^n$, have zero coefficients. Also we shall call $a_n x^n$ as the *leading term* and $a_n x^0$ is called the *constant term* and a_0 is called the zeroth coefficient of f (x). For example $f(x) = 2x^0 + 3x - 4x^2 + 4x^3 - 8x^4$ is a polynomial of degree 4 over the ring of integers. Here –8 is the leading coefficient and 2 is the zeroth coefficient. The coefficients of all terms which contain powers of x greater than 4 will be regarded as zero. Similarly $g(x) = 3x^0$ is a polynomial of degree zero over the ring of integers. In this polynomial the coefficients of x, x^2, x^3,... are all equal to zero. The zero. The zero polynomial over an arbitrary ring R will be represented by $0x^0$.

Set of constant polynomials over a ring. Let R be an arbitrary ring and R [x] the set of all polynomials over r. Let R' denote the set of all polynomials over R whose coefficients are all zero except for the constant term, which may be either zero or non-zero. That is,

$$R' = \{ax^0 : a \in R\}.$$

Then R' will be called as the set of constant polynomials in R [x].

Thus all the polynomials of degree 0 as well as the zero polynomial will be called as constant polynomials.

Example 1:

Add and multiply the following polynomials over the ring of integers:

$$f(x) = 2x^0 + 5x + 3x^2 - 4x^2,\ g(x) = 3x^0 + 4x - x^3 + 5x^4.$$

Solution:

By our definition of the sum of two polynomials, we have

$$f(x) + g(x) = (2 + 3)x^0 + (5 + 4)x + (3 + 0)x^3 + (-4 - 1)x^3 + (0 + 5)x^4$$

$$= 5x^0 + 9x + 3x^2 - 5x^3 + 5x^4.$$

Also $f(x)\,g(x) = (2x^0 + 5x + 3x^2 - 4x^3)(3x^0 + 4x - x^3 + 5x^4)$

$$= 6x^0 + (8 + 15)x + (20 + 9)x^2 + (-2 + 12 - 12)x^3 + (10 - 5 - 16)x^4 + (25 - 3)x^5 + (15 + 4)x^6 - 20x^7$$

$$= 6x^0 + 23x + 29x^2 - 2x^3 - 11x^4 + 22x^5 + 19x^6 - 20x^7.$$

Example 2:

Add and multiply the following polynomials over the ring $(I_6, +_6, \times_6)$:

$$f(x) = 2x^0 + 5x + 3x^2,\ g(x) = 1x^0 + 4x + 2x^3.$$

Solution:

$$f(x) + g(x) = (2 +_6 1)x^0 + (5 +_6 4)x + (3 +_6 0)x^2 + (0 +_6 2)x^3$$

$= 3x^0 + 3x + 3x^2 + 2x^3.$

Also $f(x)\, g(x) = (2x^0 + 5x + 3x^2)(1x^0 + 4x + 2x^3)$

$= (2 \times_6 1)\, x^0 + [(2 \times_6 4) +_6 (5 \times_6 1)]\, x + [(5 \times_6 4) +_6 (3 \times_6 1)]\, x^2$

$+ [(2 \times_6 2) +_6 (3 \times_6 4)]\, x^3 + (5 \times_6 2)\, x^4 + (3 \times_6 2)\, x^5$

$= 2x^0 + (2 +_6 5)\, x + (2 +_6 3)\, x^2 + (4 +_6 0)\, x^3 = 4x^4 + 0x^5$

$= 2x^0 + 1x + 5x^2 + 4x^2 + 4x^4.$

Note: Here degree of f (x) = 2, degree of g (x) = 3 and degree of f (x) g (x) = 4. The point to note is that degree f (x) g (x) may be less than the sum of the degrees of f (x) and g (x).

Example 3:

Add and multiply the following polynomials over the ring $(I_5, +_5, \times_5)$:

$f(x) + g(x) = 4 + 2x + x^2 + 2x^3.$

Solution:

$f(x) + g(x) = 4 + 2x + x^2 + 2x^3,$

$f(x)\, g(x) = 3 + 3x + x^2 = 3x^3 + x^4 + 4x^5.$

3.21 DEGREE OF THE SUM AND THE PRODUCT OF TWO POLYNOMIALS

Theorem:

Let f (x) and g (x) be two non-zero polynomials over an arbitrary ring R. Then

1. deg [f (x) + g (x)] ≤ Max [deg f (x), deg g (x)], if f (x) g (x) ≠ 0.
2. deg [f (x) + g (x)] ≤ deg f (x), deg g (x) if f (x) g (x) ≠ 0.

Proof:

Let $f(x) = a_0x^0 + a_1x + a_2x^2 + \dots + a_nx^n,\ a_n \neq 0$

and $g(x) = b_0x^0 + b_1x + b_2x^2 + \dots + b_mx^m,\ a_m \neq 0$

be two element of R [x].

Here deg f (x) = n and deg g (x) = m.

From our definition of the sum of two polynomials, it is obvious that if f (x) + g (x) ≠ 0, then

$$\deg [f(x) + g(x)] = \begin{cases} \max(n, m) \text{ if } \neq m \\ n \text{ if } = m \text{ and } a_n + b_m \neq 0 \\ < n \text{ if } n = m \text{ and } a_n + b_m = 0. \end{cases}$$

Again f (x) g (x) = (a_0b_0) x^0 + $(a_0b_0 + a_1b_0)$ x +...+a_nb_m x^{n+m}.

Suppose f (x) g (x) ≠ 0. Then f (x) g (x) has a unique degree.

If $a_nb_m \neq 0$, then deg [f (x) g (x)] = n + m deg f (x) + deg g (x).

Also if $a_nb_m = 0$, then deg [f (x) g (x)] < n + m.

Cor. 1: *Suppose D is an integral domain and f (x) g (x) are two non-zero elements of D [x]. Then*

deg [f (x) g (x)] = deg f (x) + deg g (x).

Proof:

Since $a_n \neq 0$, $b_m \neq 0$, therefore $a_nb_m \neq 0$ because in an integral domain the product of two non-zero elements cannot be zero. Hence deg [f (x) g (x)] = m + n.

Cor 2: *If F is afield and f (x), g (x) are two non-zero elements of F [x], then deg [f (x) g (x)] = deg f (x) = deg g (x).*

Proof:

Since a field is also free from zero divisors, therefore $a_nb_m \neq 0$ when $a_n \neq 0$ and $b_m \neq 0$. Hence the result.

3.22 RING OF POLYNOMIALS

Theorem 1:

The set R [x] of all polynomials over an arbitrary ring R is a ring with respect to addition and multiplication of polynomials.

Proof:

Let f (x), g (x) ∈ R [x]. Then f (x) + g (x) and f (x) g (x) are also polynomials over R. Therefore R [x] is closed with respect to addition and multiplication of polynomials.

Now let

f (x) = $\Sigma\, a_1x^i = a_0x^0 + a_1x + z_2x^2 + \ldots$, g(x) = $b_0x^0 + b_1x + b_2x^2 + \ldots$,

h (x) = $c_0x^0 + c_1x + c_2x^2 + \ldots$ be any arbitrary elements of R[x].

Commutativity of addition. We have

f (x) + g (x) = $(a_0 + b_0)\, x^0 + (a_1 + b_1)\, x + (a_2 + b_2)\, x^2 + \ldots$

= $(b_0 + a_0)\, x^0 + (b_1 + a_1)\, x + (b_2 + a_2)\, x^2 + \ldots$ = g (x) + f (x).

Associativity of addition. We have

$[f(x) + g(x)] + h(x) = \Sigma (a_i + b_i) x^i + \Sigma c_i x^i = \Sigma [(a_i + b_i) + c_i] x^i$
$= \Sigma [(a_i + (b_i + c_i)] x^i = \Sigma a_i x^i + \Sigma (b_i + c_i) x^i = f(x) + [g(x) + h(x)].$

Existence of additive identity. Let 0(x) be the zero polynomial over R *i.e.*, $0(x) = 0x^0 + 0x + 0x^2 + ...$

Then $f(x) + 0(x) = (a_0 + 0) x^0 + (a_1 + 0) x + (a_2 + 0) x^2 + ...$

$$= a_0 x^0 + a_1 x + a_2 x^2 + ... = f(x).$$

∴ the zero polynomial 0 (x) is the additive identity.

Existence of additive inverse. Let –f (x) be the polynomial over R defined as $-f(x) = (-a_0) x^0 + (-a_1) x + (-a_2) x^2 + ...$

Then $-f(x) + f(x) = (-a_0 + a_0) x^0 + (-a_1 + a_1) x + (-a_2 + a_2) x^2 + ... = 0x^0 + 0x + 0x^2 + ... = 0(x)$ = the additive identity.

∴ each member of R [x] possesses additive inverse.

Associativity of Multiplication. We have

$f(x) g(x) = (a_0 x^0 + a_1 x + a_2 x^2 + ...) (b_0 x^0 + b_1 x + b_2 x^2 + ...)$

$$= d_0 x^0 + d_1 x + d_2 x^2 + ... d_i x^i + ..., \text{ where } d_i = \Sigma a_i b_{j,\ i+j-i}$$

Now $[f(x) g(x)] h(x)$

$$= (d_0 x^0 + d_1 x + d_2 x^2 + ...) (c_0 x^0 + c_1 x + c_2 x^2 + ...)$$

$$= e_0 x^0 + e_1 x + e_2 x^2 + ... + e_n x^n + ...,$$

Where e_n = the coeff. of x^n in $[f(x) g(x)] h(x)$

$$= \sum_{i+k=n} d_i c_k = \sum_{i+k-n} [(\sum_{i+j=i} a_i b_j\ c_k)] = \sum_{i+j+k-n} a_i b_j c_k.$$

Similarly we can show that the coeff. of x^n in

$$f(x) [g(x) h(x)] = \sum_{i+j+k=n} a_i b_j c_k.$$

Thus $[f(x) g(x)] h(x) = f(x) [g(x) h(x)]$ since corresponding coefficients in these two polynomials are equal.

Distributivity of Multiplication with respect to addition. We have $f(x) [g(x) + h(x)]$

$$=(x_0 x^0 + a_1 x + a_2 x^2 + ...) [(b_0 + c_0) x^0 + (b_1 + c_1) x + (b_2 + c_2) x^2 + ...].$$

If n is any non-negative integer, then the coefficient of x^n in $f(x) [g(x) + h(x)]$

$$= \sum_{i+j=n} a_j (b_j + c_j) = \sum_{i+j=n} (a_i b_j + a_i c_j) = \sum_{i+j=n} a_i b_j + \sum_{i+j=n} a_i c_j$$

= Coeff. of x^n in $f(x) g(x)$ + coeff. of x^n in $f(x) h(x)$

= Coeff. of x^n = [f (x) g (x) + f (x) h (x)].

$\therefore$ f (x) [g (x) h (x)] = f (x) g (x) + f (x) h (x).

Similarly we can prove the right distributive law.

Hence R [x] is a ring. This is called the ring of all polynomials over R. There zero element of this ring is the zero polynomial

$$0x^0 + 0x + 0x^2 + 0x^3 + ...$$

Note: In future we shall write 0 in place of the zero polynomial. If $ax^0 + 0x + 0x^2 + ...$ is any constant polynomial in R [x], the we shall simply write a in place of this polynomial.

Also in place of $a_0x^0 + a_1x + a_2x^2 + ...$, we shall write $a_0 + a_1x + a_2x^2 + ...$ If we are to multiply f (x) by a constant polynomial $ax^0 + 0x + 0x^2 + ...$, then we shall simply write of (x) in place of $(ax^0 + 0x + ...)$ f (x).

Theorem 2:

if R is an integral domain with unity element, then any unit in R [x] must already be a unit in R.

Proof:

If R is an integral domain with unity element 1, then R [x] is also an integral domain with unity element. Further the constant polynomial 1 is the unity element of R [x]. Let f (x) be a unit in R [x] *i.e.*, let f (x) be an inversible element of R [x]. Let g (x) be the inverse of f (x) in R [x]. Then

f (x) g (x) = 1.

$\Rightarrow$ deg [f (x) g (x)] = 0

[$\because$ degree of the constant polynomial 1 is 0]

$\Rightarrow$ deg f (x) + deg g (x) = 0 $\Rightarrow$ deg f (x) = 0, deg g (x) = 0

$\Rightarrow$ both f (x) and g (x) are constant polynomials in R [x].

Let f (x) = a $\in$ R and g (x) = b $\in$ R. Then ab = 1 $\Rightarrow$ a is a unit in R. Thus, any unit in R [x] must already be a unit in R.

Note: If a $\in$ R is a unit in R, then a $\in$ R [x] is also a unit in R [x]. If b is the inverse of a in R, then the constant polynomial be in the inverse of a in R [x].

Polynomials Over a Field

Theorem 3:

If F is a field, then the set F [x] of all polynomials over F is an integral domain.

Proof:

Every field is a integral domain. We shall call the set F [x] as the polynomial domain over the field F.

3.23 POLYNOMIALS OVER A INTEGRAL DOMAIN

Theorem:

If D is an integral domain, then the polynomials ring D [x] is also an integral domain.

Proof:

Let D be a commutative ring without zero divisors and with unity element 1. As proved in d [x] is also a ring. To prove that D [x] is an integral domain we should prove that (1) D [x] is commutative, (2) is without zero divisors and (3) possesses the unity element.

D [x] is commutative. Let $f(x) = a_0 + a_1x + a_2x^2 + \ldots$ and $g(x) = b_0$ $b_1x + b_2x^2 + \ldots$ be any two elements of D [x].

If n is any non-negative integer, then the coefficient of x^n in f (x) g (x)

is $= \sum_{i+j=n} a_i b_j = \sum_{i+j=n} b_j a_i$, since D is commutative

= Coefficient of x^n in g (x) f (x).

$\therefore$ f (x) g (x) f (x). Hence D [x] is commutative ring.

If is the unity element of D, then the constant polynomial $1 + 0x + 0x^2 + 0x^3 + \ldots$ is the unity element of D [x]. We have

$$[a_0 + a_1x + a_2x^2 + \ldots] [1 + 0x + 0x^2 + 0x^3 + \ldots]$$
$$= (a_01) + (a_11)\, x + (a_21)\, x^2 + \ldots = a_0 + a_1x + a_2x^2 + \ldots$$

$\therefore$ The polynomial $1 + 0x + 0x^2 + \ldots$ or simply 1 is the unity element of [x].

D [x] **is without zero divisors.** Let

$$f(x) = a_0 + a_1x + a_2x^2 + \ldots + a_mx^m,\ a_m \neq 0$$
$$g(x) = b_0 + b_1x + b_2x^2 + \ldots + b_nx^n,\ b_m \neq 0$$

be two non-zero elements of D [x].

Then f (x) cannot be a zero polynomial *i.e.*, the zero element of D [x]. The reason is that at least one coefficient of f (x) g (x) namely a_mb_n of x^{m+n} is $\neq 0$ because a_m, b_n are non-zero elements of D and D is without zero divisors.

Hence D [x] is an integral domain.

Example 1:

Consider the following polynomials over the ring

$(I_8, +_8, \times_8)$:

$f(x) = 2 + 6x + 4x^2$, $g(x) = 2x + 4x^2$, $h(x) = 2\ 4x$ *and find*

1. deg [f (x) + g (x)]
2. deg [f (x) g (x)]
3. deg [h (x) h (x)].

Solution:

1. We have $f(x) + g(x) = (2 + 6x + 4x^2) + (0 + 2x + 4x^2)$
$= (2 +_8 0) + (6 +_8 2)x + (4 +_8 4)x^2 = 2 + 0x + 0x^2 = 2.$
Thus f (x) + g (x) is a non-zero constant polynomial and so deg [f (x) + g (x)] = 0.
2. We have $f(x)\, g(x) = (2 + 6x + 4x^2)(2x + 4x^2)$
$= (2 \times_8 2)x + [(2 \times_8 4) +_8 (6 \times_8 2)]x^2$
$+ [(6 \times_8 4) +_8 (4 \times_8 2)]x^3 + (4 \times_8 4)x^4$
$= 4x + (0 +_8 4)x^2 + (0 +_8 0)x^3 + 0x^4.$
$= 4x + 4x^2 + 0x^2 + 0x^4 = 4x + 4x^2.$
$\therefore$ deg [f (x) g (x)] = 2.
3. We have $h(x)\, h(x) = (2 + 4x)(2 + 4x)$
$= (2 \times_8 2) + [(2 \times_8 4) +_8 (4 \times_8 2)]x + (4 \times_8 4)x^2$
$= 4 + (0 +_8 0)x + 0x^2 = 4 + 0x + 0x^2 = 4.$
Thus h (x) h (x) is a non-zero constant polynomial and so deg [h (x) h (x)] = 0.

Theorem:

The polynomial domain F [x] over a field F is not a field.

Proof:

In order to show that F [x] is not a field, we should show that there exists a non-zero element of F [x] which has no multiplicative inverse. Let f (x) be any member of F [x] such that deg f (x) is greater than zero. The inverse of f (x) cannot be the zero polynomial the product f (x) and the zero polynomial will be equal to the zero polynomial and not equal to the unity element of F [x] which is the polynomial 1 = 0x = $0x^2$ +... Suppose now g (x) is any non-zero polynomial. Then f being a field, we have

deg [f (x) g (x)] = deg f (x) + deg g (x) > because deg f (x) > 0 and deg g (x) ≥ 0.

The degree of the unity element of F [x] is 0. Hence f (x) g (x) cannot be equal to the unity element of F [x]. Thus f (x) does not possess multiplicative inverse.

∴ F [x] is not field.

Note: The only inversible elements of F [x] are constant polynomials excluding the zero polynomial. No member of F [x] whose degree is greater then 0 is inversible.

Example 2:

Show that id a ring R has no zero divisors, then the ring R [x] has also no zero divisors.

Solution:

It is given that a ring R has no zero divisors and we have to show that the ring R [x] has also no zero divisors.

Let $f(x) = a_0 + a_1x + a_2x^2 + ... + a_mx^m, a_m \neq 0$

and $g(x) = b_0 + b_1x + b_2x^2 + ... + b_nx^n, b_n \neq 0$

be two non-zero elements of R [x].

Then f (x) g (x) cannot be the zero polynomial *i.e.*, the zero element of R [x]. Then reason is that at least on coefficient of f (x) g (x) namely a_mb_n of x^{m+n} is ≠ 0 because a_m, b_n are non-zero elements of R and R is without zero divisors.

Thus, in R [x] the product of no two non-zero elements can be the zero element. Hence the ring R [x] has no zero divisors.

Example 3:

Prove that the relation of divisibility in an integral domain is reflexive and transitive.

Solution:

Let D be an integral domain with unity element 1.

If a ∈ D, then a is said to divide be ∈ D, if there exists an element c ∈ D such that b = ca. If a is a divisor of b, then symbolically we write a | b.

1. The relation of divisibility on D is reflexive. Let a be any element of D.

   We can write a = 1 a, where 1 ∈ D.

   ∴ a | a.

   Thus a | a, ∀ a ∈ D. Hence the relation of divisibility on D is reflexive.

2. The relation of divisibility on D is transitive *i.e.*, a | b and b | c ⇒ a | c.

   We have a | b ⇒ b = ap for some p ∈ D

   and b | c ⇒ c = bq for some q ∈ D.

   Now c = bq and b = aq ⇒ c = (ap) q

   ⇒ c = a (pq) ⇒ a | c since pq ∈ D.

   Hence the relation of divisibility on D is transitive.

4

LOGICS

4.1 INTRODUCTION

In this chapter we shall study the algebra of statements or sentences. A statement, in practice, is constructed by means of words. Also we know that a word has more than one meaning, so there is a possibility of interpreting a group of words in more than one way and thereby creating a confusion in the meaning of a statement.

We shall remove this confusion or ambiguity by developing the algebra of statements, which will in turn simplifying the complicated statements. We use symbolic language to express mathematical statements and analysis of this symbolic language is the *logic*. Study of logic is very important in discrete mathematics as it provides the theoretical basis for many areas of computer science as artificial intelligence, digital logic design etc. Here we discuss a few of the basic ideas and define some of the logical concepts that are useful in computer science.

4.2 STATEMENT

If a sentence can be judged to be true or false but not both, it is called a *statement or proposition*. In other words, a sentence for which the question, "Is it true?" or "Whether it is true or false?" has a meaning, is called a *statement*. It is important to note that *every sentence is not a statement.*

Example: *Let us consider the following sentences:*

(i) What is the time by your watch?

(ii) Girls are pretty.

(iii) $\sqrt{3}$ *is an irrational number.*

(iv) There is a life on the Mars.

(v) Buck up! boys, buck up.

(vi) How are you?

(vii) Delhi is the capital of India.

Now, we will examine each of the sentences mentioned above.

We notice that the sentences (iii), (iv) and (vii) are statements because in these cases we have answer to the question "Is it true?" In sentence (iii) "Is it true that √3 is an irrational number?" has an answer. The answer can be yes or no to such questions. Similarly in sentences (iv) and (vii), the question:

Is there life on the Mars?

Is Delhi the capital of India? have answers.

These answers can be affirmative or negative, we are not concerned with that, but the important thing is that the question "Is the sentence true?" should have an answer but it is not necessary to obtain it. Thus sentences (iii), (iv) and (vii) are *statements*. On the other hand the sentences:

Girls are pretty

What is the time by your watch

Buck up! boys, buck up

How are you?

have no answer to the question "Is the sentence true?, viz.

Is "What is time by your watch? true?

is "The girls are pretty" true?

have no meaning and, therefore, they are *not statements.*

4.3 TRUTH VALUE OF A STATEMENT

We know that every statement is a sentence, which is either true or false. The truth or falsity of a statement is called its *truth value*. We assign to the statement p, the letter **T** when p is true and the letter **F**, when it is false. Both **T** and **F** are called the *truth values* of the statement p.

4.4 EQUIVALENT STATEMENTS

If the truth values of two statements are identical, then they are equivalent, *i.e.*, if p and q are two statements, then p and q are equivalent statements if p is **T** and q is also **T** or if p is **F** and q is also **F**. The equivalent statements are denoted by p = q or p ⇔ q.

Example:

Let the statements p and q be as follows:

p: n is an even integer

q: n + 1 is an odd integer.

Now, if n is an even number, then we know that n + 1 is an odd number, therefore p ⇔ q. Hence p and q are equivalent statements.

4.5 SIMPLE STATEMENT

A statement p is said to be *simple* if it has one subject and one predicate.

4.6 COMPOUND STATEMENT

A statement is said to be *compound statement* if it is formed of two or more than two simple statements. The simple statements are called the components of the compound statement. The compound statement is formed by connecting the simple statements with the help of the words **'and'** **'or'**, **'if'** ... **'then'**.

Example:

If p and q are two statements such that

p: Gunjan is an intelligent girl

q: Gunjan will join M.C.A.

Now, p and q are two simple sentences and they can be combined in a number of ways to form a *Compound* sentences such as:

(i) Gunjan is an intelligent *girl and* she will join M.C.A.

(ii) Gunjan is intelligent *or she* will join M.C.A.

(iii) If Gunjan is intelligent *then* she will join M.C.A.

(iv) Gunjan is an intelligent *girl if and only if* she joins M.C.A.

4.7 LOGIC

We use symbolic language to express mathematical statements and adjudge the truth or falsity of these statements through valid reasoning.

The analysis of this symbolic language may be termed as logic.

We usually denote statements by small letter p, q, r etc. It is clear that for a given statement p, exactly one of the following must hold:

(i) p is true;

(ii) p is false.

4.8 LOGICAL CONNECTIVES

Negation of a Statement or Contradiction

Associated with every statement is another statement called its negation. In other words, the statement having meaning contradictory to the given statement p is called the negation or contradiction of p and is denoted by '~ p' *i.e.*

'~ p' ⇒ implies not p or negative of p.

It is important to note that the negation of a *true* statement is *false* and that of a *false* statement is *true*.

i.e., if p is true then ~ p is false
if p is false then ~ p is true

Example:

The negation of the statement

p: $\sqrt{5}$ is an irrational number

is ~ p: $\sqrt{5}$ is not an irrational number.

The following truth table expresses the relation between the truth values of the statement p and those of its negation.

p	~ p
T	F
F	T

In the above table p is the statement and ~ p is the contradiction or negation of p. The truth values of p are stated is the column under p and that of ~ p are stated in the column under ~ p.

Theorem :

Prove that for any statement p, ~ (~p) = p. (Law of double negation).

Proof:

We shall prove that the statements p and ~ (~p) are equivalent. Let us prepare the truth table for these two.

Truth Table

p	~ p	~ (~ p)
(1)	(2)	(3)
T	F	T
F	T	F

The table shows that if p is **T** ⇒ ~ (~p) is also T

and if p is **F** ⇒ ~ (~p) is also F.

Hence, columns (1) and (3) show that p and ~ (~p) are equivalent statements. Hence ~ (~p) = p.

Conjunction

If the compound statement is formed by connecting the two simple statements with the help of the word **'and'** then, ***'and'*** is called the ***conjunction*** of the statements.

Example:

If p and q are two statements such that

p: Shourya is an intelligent boy

q: Shourya will join M.B.B.S.

These two statements give rise to the compound sentence 'Shourya is an intelligent boy and he will join M.B.B.S.

In this compound sentence *and* is the conjunction for p and q.

Truth Table for p ∧ q (p and q)

If p and q denote two statements then the conjunction of p and q is denoted by p ∧ q, and is read as p and q.

The statement p ∧ q is true if and only if both p and q are true. In all other cases it is false. In other words

if p is true and q is true then p ∧ q is true,

if p is true and q is false then p ∧ q is false,

if p is false and q is true then p ∧ q is false,

if p is false and q is false then p ∧ q is false,

This can be represented by following truth table.

Truth Table for $p \wedge q$

p	q	$p \wedge q$
T	T	T
T	F	F
F	T	F
F	F	F

Disjunction $\vee$ (OR)

If the compound statement is obtained by connecting the two simple statements with the help of the word *or*, then the word *or* is called the *disjunction* of the statements. If p and q are two statements, then $p \vee q$ denotes the **disjunction** of p and q. The symbol $p \vee q$ is read as p or q.

Example:

If p stands for 'I shall purchase a book' and q stands for 'I shall purchase a pencil' then $p \vee q$ = I shall purchase a book or a pencil.

Truth Table for $p \vee q$

The statement $p \vee q$ will be true if either

(i) p is true, or

(ii) q is true, or

(iii) both p and q are true.

If both p and q are false, then the statement $p \vee q$ is false. The truth table for $p \vee q$ is as follows:

Truth Table for $p \vee q$

p	q	$p \vee q$
T	T	T
T	F	T
F	T	T
F	F	F

Implication $\Rightarrow$ or Conditional Statement

A conditional statement has two simple statements connected by the word "*if, ... then*". If p and q are two statements, the word **"if, ... then"** is symbolically written as $p \Rightarrow q$. It is read as p implies q.

Example:

A father declared, "If my son stands first then I shall buy a watch for him".

The part 'If my son stands first' is called the **antecedent** and 'I shall buy a watch for him' is called the **consequent** in the above conditional statement. If we denote antecedent by p and the consequent by q then conditional statement is expressed by writing $p \Rightarrow q$. Now there are four possibilities, namely.

(i) the son stands first and his father buys a watch for him,

(ii) the son does not stand first and the father buys him a watch,

(iii) the son stands first but the father does not buy a watch for him,

(iv) the son does not stand first and the father does not buy him a watch,

It is clear that in all the four cases the father breaks his promise only in case (iii). Therefore it is the case when the conditional is false. In all other cases if is true. Thus if p is true and q is false then $p \Rightarrow q$ is false otherwise it is true in all other cases. The truth table of $p \Rightarrow q$ is given as follows:

Truth Table for $p \Rightarrow q$

p	q	$p \Rightarrow q$
T	T	T
F	T	T
T	F	F
F	F	T

Double Implication $\Leftrightarrow$ or Biconditional or Equivalence

A biconditional contains the connective **'if and only if'** and has two conditions, thus $p \Leftrightarrow q$ is same in the meaning as $p \Rightarrow q$ and $q \Rightarrow p$.

Example:

In the example given for conditional had the father promised "I shall buy a watch for my son if and only if he stands first", he would surely have meant that he would not buy a watch if his son does not stand first. As such in addition to case (iii), case (ii) as well would go against his promise and will make the conditional a false statement. Thus the statement $p \Leftrightarrow q$ is true if p and q have the same truth values and is false if they have the opposite truth values. The truth table for $p \Leftrightarrow q$ is as given below:

Truth Table for p ⇔ q

or

(p ⇒ q) ∧ (q ⇒ p)

p	q	p ⇔ q
T	T	T
F	T	F
T	F	F
F	F	T

Logically Equivalent Statements

Two statements are said to be 'logically equivalent' if both have the identical truth values, *i.e.*, in each row of the truth table both statements must have same truth values. Logically equivalent statements are expressed by writing the symbol ≡ in between or sometimes by = if there is no confusion.

Theorem (De Morgan's Law):

If p and q are two statements, then prove that

(a) $\sim (p \wedge q) = (\sim p) \vee (\sim q)$

(b) $\sim (p \vee q) = (\sim p) \wedge (\sim q)$

Proof:

(a) Let us form the truth table:

Truth Table

p (1)	q (2)	p ∧ q (3)	~(p ∧ q) (4)	~p (5)	~q (6)	(~p) ∨ (~q) (7)
T	T	T	F	F	F	F
T	F	F	T	F	T	T
F	T	F	T	T	F	T
F	F	F	T	T	T	T

In the above table columns (4) and (7) are identical and thus the statements ~ (p ∧ q) and (~p) ∧ (~q) are equivalent statements.

Hence $\sim(p \wedge q) = (\sim p) \vee \sim(q)$

(b) Let us construct the following truth table

Truth Table

p (1)	q (2)	$p \vee q$ (3)	$\sim(p \vee q)$ (4)	$\sim p$ (5)	$\sim q$ (6)	$(\sim p) \wedge (\sim q)$ (7)
T	T	T	F	F	F	F
T	F	T	F	F	T	F
F	T	T	F	T	F	F
F	F	F	T	T	T	T

In the above table we notice that columns (4) and (7) are identical, and therefore, the statements $\sim(p \vee q)$ and $(\sim p) \wedge (\sim q)$ are equivalent statements.

Hence $\sim (p \vee q) = (\sim p) \wedge (\sim q)$.

4.9 ARGUMENT

An argument is a statement which declares that the given set of propositions $p_1, p_2, ..., p_n$ yield a new proposition Q. Then argument is expressed as

$$p_1, p_2, p_3, ..., p_n \vdash Q$$

or $$p_1 \wedge p_2 \wedge p_3 \wedge ... \wedge p_n \vdash Q$$

The symbol $\vdash$ is known as **turnstile**. The proposition $p_1, p_2, p_3, ..., p_n$ are called **'premises'** (or **assumptions**) and Q is called the **'conclusion'**.

Such an argument is true, *i.e.*, valid when Q is true, *i.e.*, when all the premises $p_1, p_2, p_3, ..., p_n$ are true. It may so happen that p_1, p_2 being true (in the truth table) but conclusion is not true, the argument is false *i.e.*, not valid.

4.10 LAWS OF ALGEBRA OF PROPOSITION

Idempotent laws	$p \vee p \equiv p$	$p \wedge p \equiv p$
Associative laws	$(p \vee q) \vee r = p \vee (q \vee r)$	$(p \wedge q) \wedge r \equiv p \wedge (q \wedge r)$
Commutative laws	$p \vee q \equiv q \vee p$	$p \wedge q \equiv q \wedge p$
Distributive laws	$p \vee (q \wedge r) \equiv (p \vee q) \wedge (p \wedge r)$	
		$p \wedge (q \vee r) \equiv (p \wedge q) \vee (p \wedge r)$
Identity laws	$p \vee t = p$	$p \wedge t \equiv p$
	$p \vee t \equiv t$	$p \wedge f \equiv f$
Complement laws	$p \vee \sim p \equiv t$	$p \wedge \sim p \equiv f$
	$\sim \sim p \equiv p$	$\sim t \equiv f, \sim f \equiv t$
De Morgan's laws	$\sim(p \vee q) \equiv \sim p \wedge \sim q$	$\sim (p \wedge q) \equiv \sim p \vee \sim q$

4.11 TAUTOLOGY AND FALLACY

Prime Statement

A statement is said to be a prime statement if it does not contain any connectives.

Tautology: A statement is said to be a tautology if its truth value is T irrespective of the truth or falsity of its prime statements. A tautology is generally denoted by t.

Fallacy or Contradiction: A statement is said to be a contradiction or a fallacy if its truth value is F irrespective of the truth or falsity of its prime statements. A fallacy is generally denoted by f.

Remark: It is obvious from above definitions that if a statement is a fallacy then its negation is a tautology and vice - versa.

Logically True Statement: A statement is logically true if it is derivable from a tautology. For example, the statement is $p \vee \sim p$ is a tautology and as such it is logically true.

SOLVED EXAMPLES

Example 1:

Let p be "He is tall" and q be "He is handsome". Write each of the following statements in symbolic form using p and q.

(i) It is false that he is short or handsome

(ii) He is tall or he is short and handsome

(iii) He is tall and handsome

(iv) It is not true that he is short or not handsome

(v) He is neither tall nor handsome

(vi) He is tall but not handsome.

Solution:

(i) $\sim(\sim p \vee q)$, (ii) $p \vee (\sim p \wedge q)$, (iii) $p \wedge q$, (iv) $\sim(\sim p \vee \sim q)$,

(v) $\sim p \wedge \sim q$, (vi) $p \wedge \sim q$.

Example 2:

Let p be "It is cold" and q be "It is raining". Give a simple verbal sentence which describes each of the following statements:

(i) $\sim p$; (ii) $p \vee q$, (iii) $q \vee p$, (iv) $q \Leftrightarrow p$, (v) $p \wedge q$, (vi) $p \Rightarrow \sim q$, (vii) $\sim\sim q$, (viii) $\sim p \wedge \sim q$, (ix) $p \Leftrightarrow q$.

Solution:

(i) It is not cold

(i) It is cold or it is raining

(ii) It is cold or it is not cold

(iii) It is raining or it is not cold

(iv) It is raining if and only if it is cold

(v) It is cold and raining

(vi) It is cold then it is not raining

(vii) It is raining

(viii) It is not cold and it is not raining

(ix) It is cold if and only if it is raining.

Example 3:

In a certain country it is found that weather follows the following rules:

If it is fine today, then it is windy tomorrow

If it is calm today, then it is hot tomorrow

If it is fine tomorrow, then it is cold tomorrow

Each day is either hot or cold, wet or fine and calm or windy. Forecast tomorrow's weather if today is fine, calm and cold.

Solution:

The argument is as follows:

(i) fine today $\Rightarrow$ windy tomorrow

today is fine, $\therefore$ tomorrow is windy

(ii) calm today $\Rightarrow$ hot tomorrow

today is calm, $\therefore$ tomorrow is hot

The truth table for $p \Rightarrow q$ and $\sim q \Rightarrow \sim p$ are same, so that proposition $p \Rightarrow q$ and $\sim q \Rightarrow \sim p$ are equivalent. In other words $p \Rightarrow q \equiv \sim q \Rightarrow \sim p$.

$\therefore$ fine tomorrow $\Rightarrow$ cold tomorrow

$\equiv$ ~(cold tomorrow) $\Rightarrow$ ~ (fine tomorrow)

$\equiv$ hot tomorrow $\Rightarrow$ wet tomorrow

(iii) hot tomorrow $\Rightarrow$ wet tomorrow

tomorrow is hot, $\therefore$ tomorrow is wet.

The forecast is hot, wet and windy.

Example 4:

Simplify the following statements:

(i) $\sim\sim p$,

(ii) $\sim(p \vee \sim q)$,

(iii) $\sim(\sim p \wedge q)$,

(iv) $\sim(\sim p \vee \sim q)$,

(v) $(p \vee q) \wedge \sim p$

Solution:

(i) $\sim\sim p = p$ $\quad$ ($\therefore$ negative of a negative is positive)

(ii) $\sim(p \vee \sim q) = \sim p \wedge \sim\sim q = \sim p \wedge q$ $\quad$ ($\because \sim\sim q = q$)

(iii) $\sim(\sim p \wedge q) = \sim\sim p \vee \sim q = p \vee \sim q$ $\quad$ ($\because \sim\sim p = p$)

(iv) $\sim(\sim p \vee \sim q) = \sim\sim p \wedge \sim\sim q = p \wedge q$

(v) $(p \vee q) \wedge \sim p = \sim p \wedge (p \vee q)$ $\quad$ (Commutative law)

$= (\sim p \wedge p) \vee (\sim p \vee q)$ $\quad$ (Distributive law)

$= f \vee (\sim p \wedge q) = \sim p \wedge q.$

Example 5:

Simplify the following statements:

(i) $p \vee (p \wedge q)$ *(ii)* $\sim(p \vee q) \wedge (\sim p \wedge q)$

Solution:

(i) $p \vee (p \vee q) = (p \wedge t) \vee (p \wedge q)$ (Identity law) $\quad$ ($\because p = p \vee t$)

$= p \wedge (t \vee q)$ $\quad$ (Distributive law)

$= p \wedge t = p$ $\quad$ ($\because t \vee q = t$)

(ii) $\sim (p \vee q) \vee (\sim p \wedge q)$

$= (\sim p \wedge \sim q) \vee (\sim p \wedge q)$ $\quad$ (De Morgan's law)

$= \sim p \wedge (\sim q \vee q)$ $\quad$ (Distributive law)

$= \sim p \wedge t = \sim p$ $\quad$ (Complement law)

Example 6:

Show that the statement $(p \wedge q) \Rightarrow p$ *is a tautology.*

Solution:

Let us prepare the truth table for the statement $(p \wedge q) \Rightarrow p$.

Truth Table

p	q	$p \wedge q$	$(p \wedge q) \Rightarrow p$
(1)	(2)	(3)	(4)
T	T	T	T
T	F	F	T
F	T	F	T
F	F	F	T

In the above table we notice that column (4) under $(p \wedge q) \Rightarrow p$ has all its entries as T.

Hence $(p \wedge q) \Rightarrow$ is a tautology.

Example 7:

Prove that the statement $[(p \Rightarrow q) \wedge (q \Rightarrow r)] \Rightarrow (p \Rightarrow r)$ is a tautology

Solution:

Let us construct the truth table for the given proposition.

Truth Table

p	q	r	$(p\Rightarrow q)$	$(q\Rightarrow r)$	$(p\Rightarrow r)$	$[p\Rightarrow r\wedge q \Rightarrow r]$	$[p\Rightarrow q\wedge q\Rightarrow r] \Rightarrow p \Rightarrow r$
(1)	(2)	(3)	(4)	(5)	(6)	(7)	(8)
T	T	T	T	T	T	T	T
T	T	F	T	F	F	F	T
T	F	T	F	T	T	F	T
T	F	F	F	T	F	F	T
F	T	T	T	T	T	T	T
F	T	F	T	F	T	F	T
F	F	T	T	T	T	T	T
F	F	F	T	T	T	T	T

Since the column (8) contains all its entries as T, therefore, the given proposition is a tautology.

Example 8:

Show by means of a truth table that $\sim(p \Rightarrow q) = p \wedge \sim q$

Solution:

Let us construct the truth table for the given proposition

$$\sim (p \Rightarrow q) = p \wedge \sim q$$

Truth Table

p	q	$p \Rightarrow q$	$\sim(p \Rightarrow q)$	$\sim q$	$p \wedge \sim q$
(1)	(2)	(3)	(4)	(5)	(6)
T	T	T	F	F	F
T	F	F	T	T	T
F	T	T	F	F	F
F	F	T	F	T	F

The given statement is valid as the truth values under column (4) and column (6) are alike. Hence $\sim (p \Rightarrow q) = p \wedge \sim q$.

Example 9:

Prove by means of a truth table

$$p \Rightarrow (q \wedge r) = (p \Rightarrow q) \wedge (p \Rightarrow r)$$

Solution:

Let us construct the truth table for the given proposition.

Truth Table

p	q	r	$q \wedge r$	$p \Rightarrow (q \wedge r)$	$p \Rightarrow q$	$p \Rightarrow r$	$(p \Rightarrow q) \wedge (p \Rightarrow r)$
(1)	(2)	(3)	(4)	(5)	(6)	(7)	(8)
T	T	T	T	T	T	T	T
T	T	F	F	F	T	F	F
T	F	T	F	F	F	T	F
T	F	F	F	F	F	F	F
F	T	T	T	T	T	T	T
F	T	F	F	T	T	T	T
F	F	T	F	T	T	T	T
F	F	F	F	T	T	T	T

Hence we notice that the truth values of columns (5) and (8) are alike, therefore the two statements are equivalent.

Hence, $p \Rightarrow (q \wedge r) = (p \Rightarrow q) \wedge (p \Rightarrow r)$

Example 10:

By means of truth table, prove that

$$p \Rightarrow q = (p \Rightarrow q) \wedge (q \Rightarrow p)$$

Solution:

Let us construct the truth table for the given proposition.

Truth Table

p	q	$p \Leftrightarrow q$	$p \Rightarrow q$	$q \Rightarrow p$	$(p \Rightarrow q) \wedge (q \Rightarrow p)$
(1)	(2)	(3)	(4)	(5)	(6)
T	T	T	T	T	F
T	F	F	F	T	F
F	T	F	T	F	F
F	F	T	T	T	T

The truth values of columns 3 and 6 are alike, so the given proposition is true.

Example 11:

Prove that proposition $\sim[p \wedge (\sim p)]$ is a tautology.

Solution:

Let us prepare the truth table for the given proposition.

Truth Table

p	$\sim p$	$p \wedge (\sim p)$	$\sim[p \wedge (\sim p)]$
(1)	(2)	(3)	(4)
T	F	F	T
F	T	F	T

Since the column (4) contains T everywhere, therefore this proposition is a tautology.

Example 12:

By means of a truth table, prove.

$$p \vee (q \wedge r) = (p \vee q) \wedge (p \vee r).$$

Solution:

Let us form the truth table for the given proposition.

Truth Table

p	q	r	q ∧ r	p ∨ (q ∧ r)	p ∨ q	p ∨ r	(p ∨ q) ∧ (p ∨ r)
(1)	(2)	(3)	(4)	(5)	(6)	(7)	(8)
T	T	T	T	T	T	T	T
T	T	F	F	T	T	T	T
T	F	T	F	T	T	T	T
T	F	F	F	T	T	T	T
F	T	T	T	T	T	T	T
F	T	F	F	F	T	F	F
F	F	T	F	F	F	T	F
F	F	F	F	F	F	F	F

Since the truth values in columns (5) and (8) are identical, therefore the corresponding compound propositions are also identical.

Example 13:

Test the validity of the following argument: "If shourya studies then he will not fail in B. Tech. examination. If he does not play, then he will study. He failed in B. Tech. Therefore he played."

Solution:

We symbolize the given statement as follows:

p: Shourya studies

q: He failed in B. Tech. Entrance Examination.

r: He plays

Now the given argument is as follows:

$$p \Rightarrow \sim q, -r \Rightarrow p, q \vdash r$$

Let us now prepare the following truth table:

Truth Table

line	p (1)	q (2)	r (3)	$\sim q$ (4)	$\sim r$ (5)	$p \Rightarrow \sim q$ (6)	$\sim r \Rightarrow p$ (7)
1.	T	T	T	F	F	F	T
2.	T	T	F	F	T	F	T
3.	T	F	T	T	F	T	T
4.	T	F	F	T	T	T	T
5.	F	T	T	F	F	T	T
6.	F	T	F	F	T	T	F
7.	F	F	T	T	F	T	T
8.	F	F	F	T	T	T	F

The premises $p \Rightarrow \sim q$, $\sim r \Rightarrow p$ and q are simultaneously only in line 5.

Thus in that case the conclusion r is also true. Hence the argument is valid.

Example 14:

Prove that compound proposition is a tautology

$$[(p \Rightarrow q) \wedge p] \Rightarrow q.$$

Solution:

Let us prepare the following truth table.

Truth Table

p (1)	q (2)	$p \Rightarrow q$ (3)	$(p \Rightarrow q) \wedge p$ (4)	$[(p \Rightarrow q) \wedge p] \Rightarrow q$ (5)
T	T	T	T	T
T	F	F	F	T
F	T	T	F	T
F	F	T	F	T

Since all the entries in the 5th column are T, therefore

$[(p \Rightarrow q) \wedge p] \Rightarrow q$ is a tautology

Example 15:

Verity whether the following statements are tautology or fallacies.

(i) $(p \wedge q) \Rightarrow (p \vee q)$

(ii) $\sim(p \vee q) \Leftrightarrow (\sim p \wedge \sim q)$

(iii) $[(p \Rightarrow q) \wedge (q \Rightarrow p)] \Leftrightarrow p \Leftrightarrow q$

(iv) $[(p \vee q) \vee r] \Leftrightarrow [(p \vee (q \vee r)]$

Solution:

Let us construct the truth tables for the given propositions:

(i) **Truth Table**

p	q	$p \wedge q$	$p \vee q$	$(p \wedge q) \Rightarrow (p \vee q)$
(1)	(2)	(3)	(4)	(5)
T	T	T	T	T
T	F	F	T	T
F	T	F	T	T
F	F	F	F	T

Since all the entries in column 5 are T, therefore it is a tautology.

(ii) **Truth Table**

			(A)			(B)	
p	q	$p \vee q$	$\sim(p \vee q)$	$\sim p$	$\sim q$	$(\sim p \wedge \sim q)$	$A \Leftrightarrow B$
(1)	(2)	(3)	(4)	(5)	(6)	(7)	(8)
T	T	T	F	F	F	F	T
T	F	T	F	F	T	F	T
F	T	T	F	T	F	F	T
F	F	F	T	T	T	T	T

Since all the entries in the column 8 are T's therefore it is a tautology.

(iii) **Truth Table**

p	q	$p \Rightarrow q$	$q \Rightarrow p$	(A) $(p \Rightarrow q) \wedge (q \Rightarrow p)$	(B) $p \Leftrightarrow q$	$A \Leftrightarrow B$
(1)	(2)	(3)	(4)	(5)	(6)	(7)
T	T	T	T	T	T	T
T	F	F	T	F	F	T
F	T	T	F	F	F	T
F	F	T	T	T	T	T

Since all the entries in column (7) are T's, therefore it is a tautology.

(iv) **Truth Table**

p	q	r	$p \vee q$	(A) $(p \vee q) \vee r$	$q \vee r$	(B) $p \vee (q \vee r)$	$A \Leftrightarrow B$
(1)	(2)	(3)	(4)	(5)	(6)	(7)	(8)
T	T	T	T	T	T	T	T
T	T	F	T	T	T	T	T
T	F	T	T	T	T	T	T
T	F	F	T	T	F	T	T
F	T	T	T	T	T	T	T
F	T	F	T	T	T	T	T
F	F	T	F	T	T	T	T
F	F	F	F	F	F	F	T

Since the columns (5) and (7) are identical so, we conclude that

$$(p \vee q) \vee r \Leftrightarrow p \vee (q \vee r)$$

Since all the entries in column 8 are T's so the given statement is a tautology.

Example 16:

Verify whether the following statements are tautology or fallacies.

(i) $[(p \Rightarrow q) \wedge (q \Rightarrow r)] \Rightarrow (p \Rightarrow r)$

(ii) $p \Rightarrow [(q \vee r) \wedge \sim (p \Rightarrow \sim r)$

(iii) $(p \Rightarrow q) \Rightarrow [(q \Rightarrow r) \Rightarrow (p \Rightarrow q)]$

Solution:

Let us construct the truth tables for all the given propositions.

(i) **Truth Table**

p	q	r	$p \Rightarrow q$	$q \Rightarrow r$	(X) $(p \Rightarrow q) \wedge (q \Rightarrow r)$	(Y) $p \Rightarrow r$	$X \Rightarrow Y$
(1)	(2)	(3)	(4)	(5)	(6)	(7)	(8)
T	T	T	T	T	T	T	T
T	T	F	T	F	F	F	T
T	F	T	F	T	F	F	T
T	F	F	F	T	F	F	T
F	T	T	T	T	T	T	T
F	T	F	T	F	T	T	T
F	F	T	T	T	T	T	T
F	F	F	T	T	T	T	T

Since all the entries in column (8) are T's, so it is a tautology.

(ii) **Truth Table**

p	q	r	(X) $q \vee r$	$\sim r$	(Y) $p \Leftrightarrow \sim r$	$\sim(p \Leftrightarrow \sim r)$	$x \wedge y$	$p \Leftrightarrow x \wedge y$
(1)	(2)	(3)	(4)	(5)	(6)	(7)	(8)	(9)
T	T	T	T	F	F	T	T	T
T	T	F	T	T	T	F	F	F
T	F	T	T	F	F	T	T	T
T	F	F	F	T	F	T	T	T
F	T	T	T	F	F	T	T	T
F	T	F	T	T	T	F	F	T
F	F	T	T	F	F	T	T	T
F	F	F	F	T	F	T	T	T

Since all the entries in the column (9) are not T's, so the given statement is not a tautology.

(iii) **Truth Table**

p	q	r	(X) p ⇒ q	(Y) q ⇒ r	Y ⇒ X	X⇒(Y⇒X)
(1)	(2)	(3)	(4)	(5)	(6)	(7)
T	T	T	T	T	T	T
T	T	F	F	F	T	T
T	F	T	F	T	F	T
T	F	F	F	T	F	T
F	T	T	T	T	T	T
F	T	F	T	F	T	T
F	F	T	T	T	T	T
F	F	F	T	T	T	T

Since all the entries in the column 7 are T's, so the given statement is a tautology.

Example 17:

By means of a truth table show that proposition (p ∧ q) ∧ ~(p ∨ q) is a contradiction.

Solution:

Let us construct the truth table for the given proposition.

Truth Table

p	q	p ∧ q	p ∨ q	~(p ∨ q)	(p ∧ q) ∧ ~(p ∨ q)
(1)	(2)	(3)	(4)	(5)	(6)
T	T	T	T	F	F
T	F	F	T	F	F
F	T	F	T	F	F
F	F	F	F	T	F

Since all the entries in the column (6) are F, so the given proposition is a contradiction.

Example 18:

By means of truth table, show that

$$p \to (q \vee r) = (p \to q) \wedge (p \to r)$$

Solution:

Let us prepare the truth table for the given proposition.

Truth Table

				(A)		(B)	
p	q	r	q ∨ r	p→(q∨r)	p→q	p→r	(p→q)∨(p→r)
(1)	(2)	(3)	(4)	(5)	(6)	(7)	(8)
T	T	T	T	T	T	T	T
T	T	F	T	T	T	F	T
T	F	T	T	T	F	T	T
T	F	F	F	F	F	F	F
F	T	T	T	T	T	T	T
F	T	F	T	T	T	T	T
F	F	T	T	T	T	T	T
F	F	F	F	T	T	T	T

Since all the entries in columns (5) and (8) are identical, therefore

$$p \to (q \vee r) = (p \to q) \vee (p \to r).$$

4.12 MATHEMATICAL SYSTEM

A mathematical system consists of:

(i) ***A set or universe U.***

(ii) ***Definitions:*** *Sentences that explain the meaning of concepts that relate to the universe are known as definitions. Any term used in describing the universe itself is said to be undefined. All definitions are given in terms of these undefined concepts of objects.*

(iii) ***Axioms:*** *Assertions about the properties of the universe and rules for creating and justifying more assertions. These rules always include the system of logic.*

(iv) ***Theorem:*** *the additional assertions mentioned above.*

Example:

In the logical system, the universe consists of propositions. The axioms are the truth tables for the logical operators and the key definitions are those of equivalence and implication.

Theorem:

A true proposition derived from axioms of mathematical system is called a theorem.

Proof:

All the theorems can be expressed in terms of a finite number of propositions, p_1, p_2, ..., p_n, called the **premises,** and a proposition C, called the **conclusion**. These theorems take the form

$$p_1 \wedge p_2 \wedge \ldots \wedge p_n \Rightarrow C$$

or more simply,

$$p_1, p_2, \ldots, \text{ and } p_n \text{ imply } C.$$

When a theorem is stated, it is assumed that the axioms of the system are true. In addition, any previously proven theorem can be considered an extension of the axioms and can be used in demostrating that the new theorem is true. When the proof is complete, the new theorem can be used to prove subsequent theorems.

4.13 PROOFS

A proof of a theorem is a finite sequence of logically valid steps that demostrate that the premises of a theorem imply the conclusion.

Proofs in Propositional Calculus (Logical System)

As deseribed earlier, a theorem is a proposition that can be proved to be true. An argument that establishes truth of a theorem is called a proof.

Valid Arguments

An argument is a sequence of statements. All statements but then final one are called premises (or assumption or hypothesis). The final statement is called conclusion.

An argument is said to be logically valid, if and only if the conjunction of the premises implies the conclusion. This means that if the premises are all true, the conclusion must also be true. However, if one or more of the premises is false, so that the conjunction of all the premises is false, then the conclusion may be either true or false.

To test the validity of an argument, the following procedure may be adopted:

(i) Identify the premises and conclusion of the argument.

(ii) Construct a truth table showing the truth values of all premises and the conclusion.

(iii) Find the rows (known as *critical rows*) in which all the premises are true.

(iv) In each critical row, determine. Whether the conclusion of the argument is also true.

(a) If in each critical row the conclusion is also true, then the argument from is valid.

(b) If there is at least one critical row in which the conclusion is false, the argument form is invalid.

Rules of Inference

The rules of inference are criteria for determining the validity of an argument. Any conclusion which is arrived by following the rules of inference is called a valid conclusion, and the argument is called valid argument. Generally, for a proof we use two fundamental rules of inference.

Rule 1: *If the statement p is assumed as true and also the statement p → q is accepted as true, then q must be true.*

Symbolically,

$$\begin{array}{l} p \to q \\ p \\ \hline \therefore q \end{array}$$

In this presentation, the assertions above the horizontal line are the **premises** or **hypotheses** while the assertion below the line is the **conclusion**. The rule depicted is known as **modus ponens** or the **rule of detachment**. The validity of the argument can also seen from the truth table. For this we construct a truth table for the premises and conclusion.

Truth Table

		Premises	Conclusion	
p	q	$p \Rightarrow q$	p	q
T	T	T	T	T
T	F	F	T	F
F	T	T	F	T
F	F	T	F	F

It is clear from the truth table that there is only one case in which both premises are true (first case), and that in this case the conclusion is also true. Hence the argument is valid.

Another way of stating that the above argument is valid is that $[(p \Rightarrow q) \wedge p] \Rightarrow q$ is tautology.

Rule 2: *Whenever the two implications $p \Rightarrow q$ and $q \Rightarrow r$ are accepted as true then the implication $p \Rightarrow r$ is accepted as true.*

Symbolically:

$$\begin{array}{l} p \Rightarrow q \\ \underline{q \Rightarrow r.} \\ \therefore\ p \Rightarrow r \end{array}$$

This argument is known as a **hypothetical syllogism.**

Truth Table

p	q	r	$p \Rightarrow q$	$q \Rightarrow r$	$p \Rightarrow r$
T	T	T	T	T	T
T	T	F	T	F	F
T	F	T	F	T	T
T	F	F	F	T	F
F	T	T	T	T	T
F	T	F	T	F	T
F	F	T	T	T	T
F	F	F	T	T	T

It is clear from the first, fifth, seventh and eighth rows of the truth table that both premises are true. Since in each case the conclusion is also true, the argument is valid.

This rule may also be described as:

$$(p \Rightarrow q) \wedge (q \Rightarrow r) \Rightarrow (p \Rightarrow r)$$

is a tautology.

Additional Valid Argument Forms

There are other valid inferences. Some of them are:

Modus Tollens

The argument of the from

$$\begin{array}{r} p \Rightarrow q \\ \sim q \\ \hline \therefore \sim p \end{array}$$

This argument is called modus tollens which means "method of denying". It can easily be established by using a truth table.

Addition

The following argument form is valid.

$$\begin{array}{r} p \\ \hline \therefore \; p \vee q \end{array}$$

This form is used for making generalizations. If p is true, then more generally, p or q is true for any other statement q.

Disjunctive Syllogism

The following statement form is valid

$$\begin{array}{r} p \vee q \\ \sim q \\ \hline \therefore \; p \end{array}$$

According to this argument, when there are two possibilities and one can rule one out, the other must be the case.

4.14 METHODS OF PROOF

Direct Proof

A direct proof is a proof in which the truth of the premises of a theorem are shown to directly imply the truth of the theorem's conclusion.

Example:

For example the direct proof of the theorem: $p \rightarrow r,\ q \rightarrow s,\ p \vee q \Rightarrow s \vee r$ *is*

Step	Proposition	Justification
(1)	$p \vee q$	Premise
(2)	$\sim p \rightarrow q$	(1) conditional rule
(3)	$q \rightarrow s$	premise
(4)	$\sim p \rightarrow s$	(2), (3), chain rule
(5)	$\sim s \rightarrow p$	(4), conditional rule
(6)	$p \rightarrow r$	premise
(7)	$\sim s \rightarrow r$	(5), (6), chain rule
(8)	$s \vee r$	(7), conditional rule

Indirect Proofs

Consider a theorem P ⇒ C, where P represents $p_1 \wedge p_2 \wedge ..., \wedge p_n$, the premises. The method of indirect proof is based an the equivalence P → C ⇒ ~ (P ∧ ~C).

This logical law states that if P ⇒ C, then P ∧ ~C is always false, *i.e.*, P ∧ ~C is a contradiction. This means that a valid method of proof is to negate the conclusion of a theorem and add this negation to the premises. If a contradiction can be implied from this set of propositions, the proof is complete. Indirect proofs can often more convenient than direct proofs.

Conditional Conclusion

The conclusion of a theorem is often a conditional proposition. The condition of the conclusion can be included as a premise in the proof of the theorem. The object of the proof is then to prove the consequence of the conclusion. This rule is justified by the logical law

$$p \to (h \to c) \Leftrightarrow (p \wedge h) \to C$$

4.15 NORMAL FORMS

By comparing truth tables, we can easily conclude whether two logical expressions P and Q are equivalent. But if the number of variables increases, the process becomes tedious. For example.

a, a → b, b → c, ..., x → y, y → z ⇒ z is a theorem in propositional calculus. However, suppose that we wrote such a program and we had to write the truth table for

$$(a \wedge (a \to b) \wedge ... \wedge (y \to z)) \to z.$$

The truth table will have 2^{26} cases. At one thousand cases per second, it would take approximately 18 hours to verify the theorem.

A better method is to transform the expressions P and Q to some standard forms of expressions P' and Q' such that a simple comparison of P' and Q' shows whether P and Q are equivalent. The standard forms are called **normal forms** or **cononical forms**. There are two types of normal forms:

Dijunctive Normal Forms

In a logical expression, a product of the variables and their negations is called an **elementary product**. For example p ∧ ~q, ~p ∧ ~q, ~p ∧ q are elementary products. A sum of the variables and their negations is called an **elementary sum**. For example ~p ∨ q, ~p ∨ ~q, p ∨ ~q are elementary sums. The elementary sums or products satisfy the following properties:

(i) An elementary sum is identically true if an only if it contains at least one pair of factors in which one is the negation of the other.

(ii) An elementary product is identically false if and if it contains at least one pair of factors in which one is negation of the other.

A logical expression is said to be in disjunctive normal form if it is the sum of elementary products. For example, $p \vee (q \wedge r)$ and $p \vee (\sim q \wedge r)$ are in disjunctive normal form.

Conjunctive Normal Form

A logical expression is said to be in conjunctive normal form if it consists of a product of elementary sum.

4.16 PROPOSITIONS OVER A UNIVERSE

Let U *be a non-empty set. A proposition over* U *is a sentence that contains a variable that can take on any value in* U *and which has a definite truth value as a result of any such substitution.*

Truth Set: If p(n) is a proposition over U, the truth set of p(n) is $T_{p(n)} = \{a \in U \mid p(a) \text{ is true}\}$.

Example: *The truth set of the proposition* $\{a, b\} \cap A = \phi$ *taken as a proposition over the power set of* $\{a, b, c, d\}$ *is* $\{\phi, \{c\}, \{d\}, \{c, d\}\}$.

Tautology and Contradiction: *A proposition over* U *is a tautology if its truth set is* U. *It is a contradiction if its truth set is empty.*

Example: $(p + 1)(p - 1) = p^2 - 1$ *is a tautology over the rationals,* $x^2 - 3 = 0$ *is a contradiction over the rationals.*

Equivalence: Two propositions are equivalent if $p \Rightarrow q$ is a tautology. In terms of truth sets, this means that p and q are equivalent if $T_p = T_q$.

Example: $n + 3 = 7$ *and* $n = 4$ *are equivalent propositions over the integers.*

Implification: If p and q are propositions over U, p implies q if $p \rightarrow q$ is a tautology. Since the truth set of $p \rightarrow q = T_p' \cap T_q$, then $p \Rightarrow q$. When $T_p \subset T_q$.

Example: Over the natural numbers: $n \leq 3 \Rightarrow n \leq 6$ since $\{0, 1, 2, 3,\} \subset \{0, 1, 2, 3, 4, 5, 6,\}$.

Example 19:

Represent the argument.

If shourya studies hard, then he gets first division.

Shourya studied hard. ... He got first division, symbolically and determine whether the argument is valid.

Solution:

Let

p: shourya studies hard,

q: shourya gets first division

The argument may be written symbolically as

$$\begin{array}{l} p \Rightarrow q \\ p \\ \hline \therefore q \end{array}$$

Hence, by modus ponens the argument is valid.

Example 20:

Represent the argument.

If it rains today, then we will not play cricket today.

If we don't play cricket today, then we will play cricket tomorrow. ...

Therefore, if it rains today, then we will play cricket tomorrow,

symbolically and determine whether the argument is valid.

Solution:

Let

p: It is raining today

q: We will not play cricket today

r: We will play cricket tomorrow.

The argument is of the form

$$\begin{array}{l} p \Rightarrow q \\ q \Rightarrow r \\ \hline \therefore \quad p \Rightarrow r \end{array}$$

Hence the argument is a hypothetical syllogism and thus the argument is valid.

Example 21:

Represent the argument.

If this number is divisible by 4, then it is divisible by 2.

This number is not divisible by 2. ∴

This number is not divisible by 4.

symbolically and determine whether the argument is valid.

Solution:

Let

p: The number is divisible by 4.

q: It is divisible by 2.

The argument may be written as

$$\begin{array}{l} p \Rightarrow q \\ \quad \sim q \\ \hline \therefore \quad \sim p \end{array}$$

Thus by modus tollens the argument is valid

Example 22:

Represent the argument.

Either Gunjan is not guilty or Rashmi is telling the truth. ...

Rashimi is not telling the truths

Therefore Gunjan is not guilty.

symbolically and determine whether the argument is valid.

Solution:

Let

p: Gunjan is not guilty

q: Rashmi is telling the truth

The argument can be written as

$$\begin{array}{l} p \vee q \\ \quad \sim q \\ \hline \therefore \quad p \end{array}$$

Thus by disjunctive syllogism, the argument is valid.

Example 23:

Prove that s is a valid conclusion from the premises $p \Rightarrow q$, $p \Rightarrow r$, $\sim (q \wedge r)$ *ad* $s \vee p$.

Solution:

To prove the theorem we have the following table:

1.	$p \Rightarrow q$	Premise (given)
2.	$p \Rightarrow r$	Premise (given)
3.	$(p \Rightarrow q) \wedge (p \Rightarrow r)$	From 1 and 2
4.	$\sim(q \wedge r)$	Premise given)
5.	$\sim q \vee \sim r$	De Morgan's law from 4
6.	$\sim p \vee \sim p$	Using 3 and 5
7.	$\sim p$	Idempotent law from 6
8.	$s \vee p$	Premise (given)
9.	s	Disjunctive syllogism from 7 and 8

Thus s is valid from the given premises.

Example 24:

Prove the validity of the following argument "If I get the admission and work hard then I will get first division. If I get first division, then I will be happy. I will not be happy.

Therefore, either I will not get the admission or I will not work hard".

Solution:

Let

p: I get the admission

q: I work hard

q: I get first division

s. I will be happy.

Then the above argument can be written in symbolic for as

$$(p \wedge q) \Rightarrow r$$
$$r \Rightarrow s$$
$$\sim s$$

Therefore,

1.	$(p \wedge q) \Rightarrow r$	Premise (given)
2.	$r \Rightarrow s$	Premise (given)
3.	$(p \wedge q) \Rightarrow s$	Hypothetical syllogism by 1 and 2
4.	$\sim s$	Premise (given)

5. $\sim(p \wedge q)$ Modus tollens by 3 and 4
6. $\sim p \vee \sim q$ Conclusion.

Hence the argument is valid.

Example 25:

Prove that product of two odd integers is an odd integer.

Solution:

Let p and q are two odd integers. Then there exist two integers m and n, so that $p = 2m + 1$ and $q = 2n + 1$. Then

$$pq = (2m + 1)(2n + 1) = 4mn + 2m + 2n + 1$$
$$= 2(2mn + m + n) + 1 \text{ which is odd.}$$

Example 26:

Prove that $\sqrt{3}$ is irrational by giving a proof by contradiction.

Solution:

Let $\sqrt{3}$ is a rational. Under assumption that $\sqrt{3}$ is rational, there exist integers p and q such that $\sqrt{3} = p/q$. Where p and q have no common factors. Squaring both sides, we get

$$3 = p^2/q^2 \Rightarrow 3q^2 = p^2$$

Hence p^2 is a multiple of 3, and therefore odd. This implies p is even. Hence $p = 3k$ for some integer k. Then $3q^2 = (3k)^2 \Rightarrow q^2 = 3k^2$. Thus q^2 is odd, q is odd. But now p and q have a common factor of 3, which is a contradiction to the statement that p and q have no common factors.

Hence our initial assumption that $\sqrt{3}$ is rational is false. Thus $\sqrt{3}$ is irrational.

Example 27:

Prove that for every positive integer n, $n^3 + n$ is even.

Solution:

When n is even: Then $n = 2k$ for some positive integer k.

Now $n^3 + n = (2k)^3 + 2k = 8k^3 + 2k$

$= 2(4k^3 + k)$ which is even.

When n is odd: Then $n = 2k + 1$ for some positive integer k.

Now $n^3 + n = (8k^3 + 12k^2 + 6k + 1) + (2k + 1)$

$= 8k^3 + 12k^2 + 8k + 2$

$= 2(4k^3 + 6k^2 + 4k + 1)$

Which is even

Hence the sum $n^3 + n$ is even.

Example 28:

Show that the following argument is invalid:

If Vipin solved this problem, then he obtained the answer 7

Vipin obtained the answer 7

Therefore, Vipin solved this problem correctly.

Solution:

Let

p: Vipin solved this problem

q: Vipin obtained the answer 7.

Then this argument is of the form: if $p \Rightarrow q$ and q, then p. This argument is faulty because the conclusion can be false even though $p \Rightarrow q$ and q are true. That is, in the implication $[(p \Rightarrow q) \wedge q] \Rightarrow p$ is not a tautology. It is possible, Vipin obtained the correct answer 7 by luck, guessing or prior knowledge but the arguments and intermediate steps are wrong. Hence the argument is invalid.

Example 29:

Obtain the disjunctive normal forms of the following:

(a) $p \vee (\sim p) \Rightarrow (q \vee (q \Rightarrow \sim r)))$

(b) $p \Rightarrow (p \Rightarrow q)\ [(\vee \sim(\sim q \vee \sim p)]$

Solution:

(a) $p \vee (\sim p) \Rightarrow (q \vee (q \Rightarrow \sim r)))$

$$\equiv p \vee (\sim p \Rightarrow (q \vee (\sim q \vee \sim r)))$$

$$= p \vee p \vee q \vee \sim q \vee -r$$

$$\equiv p \vee q \vee \sim q \vee r$$

Which is the required disjunctive normal form.

(b) $p \Rightarrow ((p \Rightarrow q) \wedge \sim (\sim q \vee \sim p)$

$$\equiv \sim p \vee ((p \Rightarrow q \wedge \sim (\sim q \vee -p))$$

$$\equiv \sim p \vee ((\sim p \vee q) \wedge \sim (\sim q \vee -p))$$

$$\equiv \sim p \vee ((\sim p \vee q) \wedge (q \wedge p)$$

$$\equiv \sim p \vee [(\sim p \wedge (q \wedge p)) \vee (q \wedge p))]$$

$$\equiv \sim p \vee [(\sim p \wedge \sim p \sim q)] \vee (q \wedge p)$$
$$\equiv \sim p \vee (p \wedge q)$$

Which is the required disjunctive normal form.

Example 30:

Obtain a conjunctive normal from of

$$[q \vee (p \wedge q)] \wedge \sim [(p \vee r) \wedge q]$$

Solution:

$$[q \vee (p \wedge q)] \wedge \sim [(p \vee r) \wedge q]$$
$$\equiv [q \vee (p \wedge r)] \wedge [\sim(p \vee r) \vee \sim q]$$
$$\equiv [q \vee (p \wedge r)] \wedge [(\sim p \wedge \sim r) \sim q]$$
$$\equiv (q \vee p) \wedge (q \vee r) \wedge (\sim q \vee \sim q) \wedge (-r \wedge \sim q)$$

This is the required conjunctive normal form.

4.17 QUANTIFIERS

If p(n) is a proposition over a universe U, its truth set $T_{p(n)}$ is equal to subset of U. In many cases, we are most concerned with whether $T_{p(n)}$ is empty or not. In other cases, we might be interested in whether $T_{p(n)} = U$, *i.e.*, p(n) is a tautology. Since the conditions $T_{p(n)} \neq \phi$ and $T_{p(n)} = U$ are very important, we have a special system of notation for them, which are called **quantifiers.**

The Existential Quantifier

If p(x) is a proposition over U with $T_{p(n)} \neq \phi$, we say "There exists an n in U such that p(n) is true". We abbreviate this sentence with the symbols $(\exists\, n)_{\cup}$ (p(n)). $\exists$ is called the **existential quantifier**. Symbol $\exists$ denotes 'there exists'.

Example:

Consider the sentence "there exists x such that $x^2 = 3$". This sentence can be written as

$$(\exists\, x \subset R)\ P'(x) \text{ or } \exists\, x\ P(x)$$

where $\quad P(x)$ "$x^2 = 3$".

This statement is called an **existential statement**.

The Universal Quantifier

If p(x) is a proposition over U with $T_{p(n)} = U$, we say "for all n in U, p(n) (is true)". We abbreviate this proposition with the symbol $(\forall\, n)_{\cup}$ (p(n)). $\forall$ is termed as the **universal quantifier** and it denotes **'for all'**.

Example:

Consider the sentence "All human beings are mortal".

Let P(x) denote "x is mortal".

Then the above sentence can be written as

$$(\forall\ x \in S)\ P(x) \text{ or } \forall\ x\ P(x)$$

This statement is called **universal statement**.

Translating Sentences into Logical Expressions

The logical operators and quantifiers can be used to express English sentences into logical expressions. Let us consider the statement.

Every bird can fly.

We translate this logically as

For every x, if x is bird then x can fly.

Using B for the bird and F for the predicate can fly, we can write

For every x, $B(x) \Rightarrow F(x)$

or $\forall\ x,\ B(x) \Rightarrow F(x)$.

Statements containing words like every, each and every one indicate universal quantifier. Sentences of this type must be reworded such that they start with for every x, which is then translate to x.

Again, let us consider the sentence

Some men are genius

We translate this as

There exist men who are genius.

Let p be the property is genius. Then the sentence can be written as $\exists$ x p(x).

Negation of Quantified Statements

When we negate a quantified proposition the existential and universal quantifiers complement one another.

Example:

Consider the statement "All students in the class have taken a course in mathematics". This statement can be written as

$$\forall\ x\ p(x)$$

Where p(x) is the statement "x has taken a course in mathematics". Its negation will be

"It is not the case that all students in the class have taken a course in mathematics".

This is equivalent to

"There is a student in the class who has not taken a course in mathematics".

This can be written as

$$\exists\, x \sim p(x)$$

Therefore, we get the following equivalence

$$\sim \forall\, x\; p(x) \equiv \exists\, x \sim p(x).$$

Thus, the negation of a universal statement is logically equivalent to an existential statement.

Similarly, we can show that the negation of a existential statement is logically equivalent to a universal statement.

Multiple Quantifiers

If a proposition has more than one variable, then we can quantify it more than once. For example, p (x, y): $(x + y)^2 = x^2 + 2xy + y^2$ is a tautology over the set of all pairs of real numbers because it is true for each pair (x, y) in **R** × **R**. Another way to look at this proposition is as a proposition with two variables.

The assertion that p (x, y) is a tautology could be quantified as

$$(\forall\, x)_{\mathbf{R}}\, (\forall\, y)_{\mathbf{R}}\, (p\,(x, y)) \text{ or}$$

$$(\forall\, y)_{\mathbf{R}}\, (\forall\, x)_{\mathbf{R}}\, (p\,(x, y))$$

In general, multiple universal quantifiers can be arranged in any order without logically changing the meaning of the resulting proposition.

The same is true for multiple existential quantifiers, for example, p (x, y): x + y = 6 and x – y = 2 is a proposition over **R** × **R**.

$$(\exists\, x)_{\mathbf{R}}\ (\exists\, y)_{\mathbf{R}}\ (x + y = 6 \text{ and } x - y = 2)$$

and $$(\exists\, y)_{\mathbf{R}}\ (\exists\, x)_{\mathbf{R}}\ (x + y = 6 \text{ and } x - y = 2)$$

are equivalent.

Thus, we have

$$\forall\, x\, \forall\, y\; p(x, y) \equiv \forall\, y\, \forall\, x\; p(x, y) \qquad \text{...(1)}$$

$$\exists\, x\, \exists\, y\; p(x, y) \equiv \exists\, y\, \exists\, x\; p(x, y) \qquad \text{...(2)}$$

Both the sides of (1) are true if and only if p(x, y) is true for all possible pairs x and y and false if there is a pair x, y for p(x, y) is false. Both sides of (2) are true if there is a single pair of x, y that makes p(x, y) true and false if p(x, y) is false for every pair x, y.

When existential and universal quantifiers are mixed, the order can not be exchanged without possibly changing the meaning of the proposition.

For example: The statement $\exists$ x $\forall$ y p(x, y) and $\forall$ y $\exists$ x p(x, y) are not logically equivalent.

Example 31:

If A = {1, 2, 3, ..., 9}. Determine the truth value of each of the following statements :

(a) $(\exists x \in A), x + 4 = 10$

(b) $(\forall x \in A), x + 4 < 15$

(c) $(\exists x \in A), x + 4 > 15.$

Solution:

(a) True, for if x = 6, then 6 + 4 = 10

(b) True, for every number is A satisfies x + 4 < 15

(c) False, for every number is A, x + 4 > 15 is false.

Example 32:

Let A(x): x is an integer, B (x): either positive or negative. Express the statement "Any integer is either positive or negative" using quantifiers.

Solution:

The given sentence can be written as

For all x, if x is an integer, then x is either positive or negative

i.e., $(\forall x) (A(x) \Rightarrow B(x))$

Example 33:

Let A(x): x is businessman, B(x): x is dishonest K(x): x is successful. Express the following using quantifiers.

(a) There exists a businessman

(b) Some businessmen are dishonest

(c) Some businessmen are not successful.

Solution:

(a) $(\exists x) (A(x))$

(b) There exists an x such that x is businessman and x is dishonest.

$(\exists x)(A(x) \wedge B(x))$

(c) There exists an x such that x is businessman and x is not successful.

$(\exists x)(A(x) \wedge \sim K(x))$.

Example 34:

Negate the statement for all real numbers x, if $x > 4$ then $x^2 > 16$.

Solution:

Let P(x) and Q(x) denote "$x > 4$" and "$x^2 > 16$"

Then the given statement can be written as

$$\forall x\, (P(x) \Rightarrow Q(x))$$

This is equivalent to

$$\forall (x\, (\sim P(x) \vee \sim Q(x))$$

The negation of this statement is

$$\exists x\, (P(x) \wedge \sim Q(x))$$

There exist a real number x such that $x > 4$ and $x^2 \leq 16$.

EXERCISES

1. If we denote the statement 'I purchased a car' by p, then write the following statements.

   (a) $\sim p$, (b) $\sim(\sim p)$

2. Let p be the statement "He is tall" and q be "He is handsome". Give a simple verbal sentence which describes each of the following statements.

   (i) $p \wedge \sim q$, (ii) $p \vee (\sim p \wedge q)$, (iii) $\sim(\sim p \vee \sim q)$.

3. Express the following statements in words:

   (a) $p \wedge q$, (b) $(\sim p) \wedge q$, (c) $(\sim p) \wedge (\sim q)$, (d) $(\sim p) \vee (\sim q)$

   Where the statements p and q are given by

   p: I shall purchase a tape recorder

   q: I shall not purchase a car.

4. If the truth values of the statements p, q, r are T, F, T respectively, then give the truth values of the following statements:

   (i) $p \vee q$,

   (ii) $p \vee (\sim r)$

(iii) $(\sim p) \vee (\sim q)$,

(iv) $(p \vee q) \wedge r$, (v) $\sim(p \vee r)$, (vi) $\sim(p) \wedge (\sim q)$

5. Construct the truth tables for:

   (a) $(\sim p) \wedge q$, (b) $\sim(p \wedge q)$

   (c) $p \vee (\sim q)$, (d) $\sim[p \vee (\sim q)]$

   (e) $(\sim p) \wedge (\sim q)$

6. Prove that the proposition $[\sim p \wedge (\sim p)]$ is a tautology.

7. Show that the statement $p \wedge \sim p$ is a fallacy.

8. Prove De Morgan's Law

   $\sim(p \wedge q) \Leftrightarrow (\sim p) \vee (\sim q)$

   Also show that the given statement is a tautology.

9. Make an appropriate truth table and test the validity of the following arguments.

   "If Piyush is a junior he is taking English. Piyush is junior, therefore, he is taking English".

10. Prove that the sentence "It is raining or it is not raining" is a tautology.

11. Establish the following results using the truth tables:

    (a) $p \vee p = p$,

    (b) $p \wedge p = p$,

    (c) $p \vee [(\sim p) \wedge q] = p \vee q$,

    (d) $(p \vee q) \wedge [p \vee (\sim q)] = p$

12. Examine the logical validity of the following either by constructing truth tables or by arguments:

    (i) $[(p \vee \sim p) \vee p] \Leftrightarrow q$,

    (ii) $[(p \vee \sim p) \Rightarrow q] \Leftrightarrow q$,

    (iii) $[(p \vee \sim p) \Leftrightarrow q] \Leftrightarrow q$,

    (iv) $[(p \wedge \sim p) \vee q] \Leftrightarrow q$,

13. By means of a truth table, prove that:

    (i) $\sim(p \Leftrightarrow q) \equiv \sim p \Leftrightarrow q \equiv p \Leftrightarrow \sim q$

    (ii) $p \vee (q \wedge r) \equiv (p \vee q) \wedge (p \vee r)$

(iii) $p \Rightarrow (q \wedge r) \equiv (p \Rightarrow q) \wedge (p \Rightarrow r)$

(iv) $(p \Leftrightarrow q) \Leftrightarrow r \equiv p \Leftrightarrow (q \Leftrightarrow r)$

(v) $p \vee q \equiv (p \downarrow q) \downarrow (p \downarrow q) \equiv \sim(p \downarrow q)$

14. Prove the following:

    (i) Conjunction or disjunction of a statement with itself is equivalent to the statement.

    (ii) the double negation of a statement is equivalent to the statement.

    (iii) a statement which implies own negation is self-contradiction.

    (iv) a logically true statement and a self-contradiction implies every statement.

15. Test the validity of:

    (a) "If the train is late, I shall miss my appointments; if it is not late, I shall miss the train; but either it will be late or not late, therefore, in any case, I shall miss my appointment.

    (b) "On my son's birthday, I bring his toys. Either it is my son's birthday or I work late in college; I did not bring my son's toys today; Therefore, today I worked late".

16. By means of truth table, prove that

$$p \vee q \equiv (p \vee q) \wedge \sim (p \wedge q)$$

17. By means of a truth table, prove that

$$p \wedge (q \vee r) \equiv (p \wedge q) \vee (p \wedge r)$$

18. A tautology is a statement which is always true. Using truth tables, show that $(p \wedge q) \Rightarrow p$ and $p \Rightarrow (p \vee q)$ are both tautologies where p, q are any two statements.

19. Show that the statement

$$p \vee (q \wedge r) \Leftrightarrow (p \vee q) \wedge (p \vee r),$$

where p, q, r are statements, is a tautology.

20. Prove that:

    (a) $p \Rightarrow q = (\sim p) \Rightarrow (\sim q)$

    (b) $\sim (p \Rightarrow q) = p \wedge (\sim q)$

    (c) $p \Rightarrow q = (\sim p) \vee q$

    (d) $[(p \rightarrow q) \wedge \sim q] \rightarrow \sim p.$

21. Show that $t \Rightarrow s$ is a valid conclusion from the given premises

    $(p \wedge q) \vee (r \Rightarrow s)$,

    $t \Rightarrow r$, $\sim (p \wedge q)$

22. Show that s is a valid conclusion from the given premises

    $p \Rightarrow \neg q$, $q \vee r$,

    $\neg s \Rightarrow p$, $\sim r$,

23. Check the validity of the following argument:

    (a) "If I study, then I will pass in examination. If I do not go to cinema, then I will study, But I failed in examination. Therefore, I went to cinema".

    (b) "If today is Monday, then Yesterday was Sunday. Yesterday was Sunday. Today is Monday".

    (c) "If I try hard and I have a talent, then I will become a scientist. If I become scientist, then I will be happy. Therefore, if I will not be happy, then I did not try hard or I do not have talent".

    (d) "If I drive to work then I will arrive in time. I do not drive to work. Therefore, I will not arrive in time".

24. Prove that $\sqrt{5}$ is irrational.

25. Prove by contradiction that the difference of any rational number and any irrational number is irrational.

26. Show that each of the following inferences is fallacy:

    (a) If the client is guilty, then he was at the scene of the crime. The client was at the scene of the crime. Hence the client is not guilty.

    (b) If today is Ankit's birthday, then today is July 18. Today is July 18. Hence today is Ankit's birthday.

27. Consider the following statement $\forall$ integers n, n^2 is odd then n is odd:

    Which of the following are equivalent ways of expressing this statement.

    (a) Given any integer whose square is odd, that integer is itself odd.

    (b) Any integer with an odd square is odd.

    (c) All integers have odd squares and are odd.

    (d) All odd integers have odd squares.

    (e) For all integers, there are some whose square is odd.

(f) If the square of an odd integer is odd, then that integer is odd.

28. Rewrite the following argument using quantifiers, variables, and predicate symbols:

    (a) There is a student who likes mathematics but not commerce

    (b) Some men are genius

    (c) Some numbers are not rational

    (d) Not all birds can fly

    (e) All birds can fly.

29. Write negations for each of the following:

    (a) $x \in R$ if $x(x+1) > 0$ then $x > 0$ or $x < -1$.

    (b) Real number x, if $x > 2$, then $x^2 > 4$

30. Show that $p \Rightarrow (s \vee t)$ is a valid conclusion from the given premises

    $p \Rightarrow (q \vee r)$,

    $q \Rightarrow s$ and

    $r \Rightarrow t$.